AF877145

Acknowledgements

I begin by thanking all the isopod hobbyists who made a hobby where there was none within a few short years of the first edition of this handbook's publishing. Special thanks go to Annie Lastar for editorial assistance this edition. Ferenc Vilisics and Christian Schwarz offered editorial suggestions for the first edition. I thank my wife, two daughters and son for their support, and God for creating varied crustaceans and my interest.

In memory of Harry Bowman whose love for plants and greenhouse taught me about humidity and moisture.

Front cover: *Cubaris* sp. "rubber ducky," pale specimens of the standard coloration.

Back cover: (top to bottom and left to right) *Porcellio werneri*, *Armadillidium klugii* "slano", *Porcellio ornatus* "high yellow", *Cubaris* sp. "Jupiter", *Armadillidium maculatum* with broken margins and peach *A. nasatum*, *Cubaris* species "Pak Chong", albino or white form of *Armadillidium vulgare*.

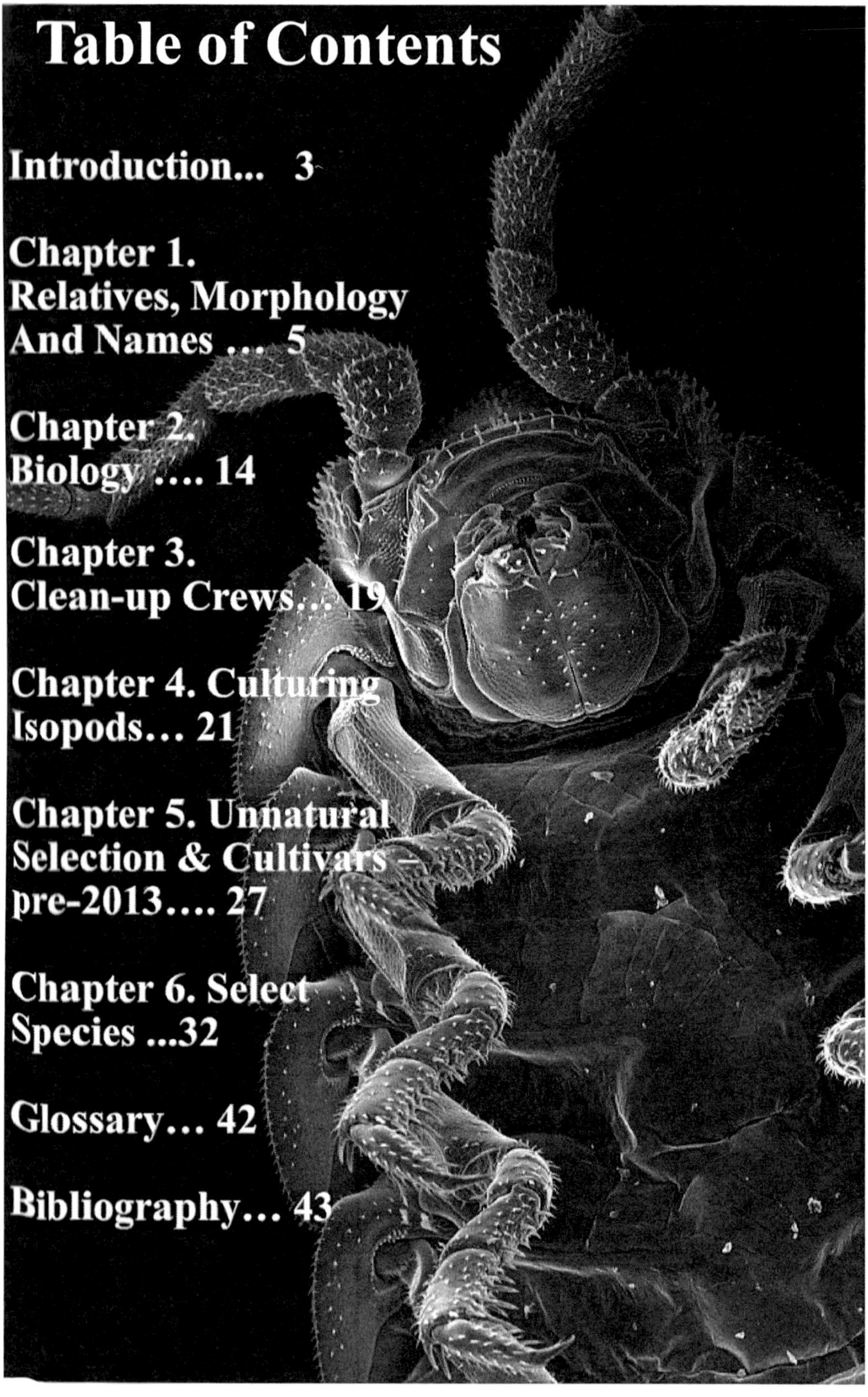

Table of Contents

Chapter 1. Relatives, Morphology & Names

The Order Isopoda is a large, diverse group of arthropods from the Class Crustacea. The Crustacea includes several shrimp-like creatures from tadpole shrimp and brine shrimp to mantis shrimp and true shrimp. The subclass Malacostraca includes the most familiar crustaceans such as crabs, lobsters, shrimp, and isopods. Freshwater crayfish, various shrimp, and a few crabs have become increasingly popular aquarium pets in recent years. Interest in crayfish and miniature shrimp somewhat resemble the isopods since much of the interest has come from rearing and developing attractive color forms under captive conditions. Orange, blue, or white crayfish forms are seen across various genera and species while the colors and patterns seen in multiple varieties of a few common, miniature shrimp are far more developed.

The malacostracan Superorder Pericarida includes the Isopoda, Amphipoda, Mysida, and some lesser-known crustaceans. Amphipods are the common scuds from marine aquaria, but they include freshwater scuds, tiny hopping terrestrials known as grass shrimp, and beach hoppers. Mysid shrimp had been a popular prey item cultured for seahorses and other picky feeders maintained in marine aquariums. Mysida are also known as opossum shrimps because, like all the Pericaridia, they hold eggs and young in a brood pouch or marsupium and the immatures are somewhat large and look like the adults (versus the nearly microscopic planktonic larvae of most true shrimp).

Within the Order Isopoda there is incredible diversity. There are more than 10,500 species known (McCormick 2007). These are distributed in eleven different suborders. The largest suborder, Oniscidea, is what we normally see, but the long-legged aquatic *Asellota* can also be encountered in many areas. Isopods are the only arthropod order to conquer virtually every habitat type on earth.

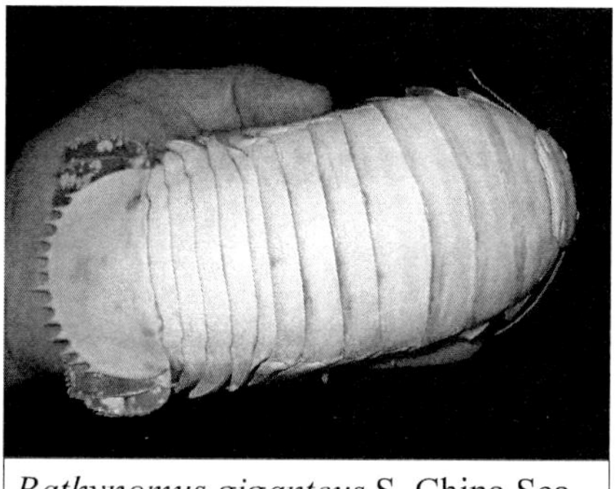

Bathynomus giganteus S. China Sea

A little over half, more than 5,000 species, are found throughout the ocean from the deepest trenches to tide pools. The largest known are among the most common of all deep sea denizens, *Bathynomus* spp. from the Suborder Cymothoida. The largest of all known isopods, *Bathynomus giganteus*, can reach an impressive 7.5 to 14.5" (19 to 37cm) though some unverified, whimsical fish stories suggest *Bathynomus* species can exceed two feet. Specimens trapped in deep waters off the coast of Australia, India, and Mexico have proven virtually identical (Caldwell 2008). It may have been the first isopod ever to make its way to television. The *Living Planet* (1984), incorrectly calls *Bathynomus* an amphipod and claims it is related to horseshoe crabs which are not crustaceans. More recently, deep sea isopods were fictionally used by NASA to fake alien fossils in the pop-fiction *Deception Point* (Brown 2001). Several public aquariums have tried to keep *Bathynomus* since the late 1980s, but little has been observed beyond short-term survival.

Various marine isopods, primarily those in the Suborder Cymothoida, colonize the bodies of fish, shrimp, and related animals as parasites for at least part of their lives. The most unusual and intriguing of all isopods is one that lives in the mouths of fish and is found in waters from the coast of California to South America. This species, *Cymothoa exigua*, parasitizese the tongue, where it lives from the fish's blood. The tongue eventually dies off due to physiological stress, which normally would doom the fish. However, *C. exigua* then takes the place of the tongue and becomes a fully functional replacement. This is the only case known in which a parasite fully replaces an organ of its host. One look in the affected fish's mouth reveals a crazy looking tongue with legs and a pair of eyes looking back at the observer. This inspired a 2012 horror flick featuring parasitic isopods as villains, but the most interesting aspect is unfortunately lost to cheap gore and poor writing.

There are some marine isopods growing as large as 12" (30 cm) that can be found in shallow waters, but these large species have not been seen in the marine aquarium hobby because they are from the frigid waters of Antarctica. A whole host of other crustaceans such as spiny lobsters, slipper lobsters, mantis shrimp, hermit crabs, etc. are popular fares in marine aquaria but these large handsome specimens can be kept near room temperature and near atmospheric pressure. A variety of very small tropical species from those with short legs and fat bodies that possess lobster-like swimming tails to those resembling house centipedes with extremely long legs and thin bodies have long been common imports on "live rock." These can be important, or at least significant, inhabitants of mini-reef aquaria (for thirty years I have seen a small species resembling the freshwater asellotids in marine pet shops and my own reefs). I had a fascinating member of the suborder Sphaeromatidea in one of my reef tanks for three or four years. It superficially resembled the familiar terrestrial species, but the males had massive, spiked uropods and though it walked like common terrestrials it would swim in rapid bursts when disturbed. A variety of species are common in sea grass beds and among marine macroalgae (seaweed). In June 2010, off a beach in Norfolk, Virginia I collected a tiny ~30mm piece of sea lettuce to see if it would grow in my reef tank. Wave action tore fragments off large plants attached to rocks somewhere offshore. I grabbed one tiny, flat blade between a finger and thumb as it floated in the water. One of the smaller pieces was chosen since it would be much easier to keep alive for a few days in transport. The next day I was shocked to discover a rather large, 28 mm x 7 mm bright green isopod from the suborder Idoteidae in the small deli-cup. This and similar species on the west coast are probably incredibly common but rarely noticed even when they wash up because they perfectly match the color of the seaweed they live on. Isopods are not just washed up on shore, they are often the dominant fauna of many shoreline habitats and some look similar to inland terrestrial species.

There are nearly 4,000 terrestrial species (Schmalfuss 2003) though there may be some debate on aquatic vs. terrestrial status for certain species found where land and sea meet. Further inland, the isopods are overshadowed by various larger life forms but are still the most common fauna in my yard and the nearby forest (with exception of microscopic or near microscopic mites, springtails, protozoans, and bacteria). Although suited to moist habitats, they can be common in New World and Old World deserts. *Venezillo arizonicus* is native to the Sonoran and Colorado deserts, the driest and hottest areas of the United States. In the Sonoran Desert, isopods are encountered in far greater numbers than desert cockroach nymphs (Polyphagidae, *Arenivaga* spp.) among mesquite leaf litter accumulations. One large species encountered in Arizona

1
2
3
4
5
6
7

(Color Plate page 7)

1. Yellow Cave Isopods, *Porcellio bolivari.*
2. The isolated orange from of *Armadillo officinalis* is brown compared to some orange forms, but there is also a geographic brown form.
3. This is the geographic form of *P. expansus* commonly called orange (right specimen is the same stock but looks like the yellow geographic form).
4. The beautiful pied "Mardis Gras" form of *Oniscus asellus* is only mostly true breeding.
5. Orange form of *Venezillo parvus* from Miami stock 2004.
6. The original Spanish orange *P. scaber* untainted from 1997 stock have never thrown a gray reversion since original isolation.
7. The isolated orange form of *Cubaris murina* is more of a fleshy pink and is traded under the name "papaya."

Armadillidium maculatum irregulars with more white are traded as "Dalmatian"

Large female *Porcellio succinctus* guarding brood of mancae molting shortly after emergence.

Compare *A. maculatum* with solid margins to high white above and back cover broken margins, same stock.

Cubaris sp. "panda king" adults are variable in the number of white-marked thoracic segments (1-5).

appears to be *Porcellio laevis* though it borrows and is highly resistant to desiccation (McMonigle 2004). *Hemilepistus* species look similar to the well-known *Porcellio* (and were considered members of the genus *Porcellio* for nearly a hundred years) but have conquered Old World deserts mostly through behavioral modifications. Terrestrial isopods are found in habitats with moisture gradients between swamp and desert, but also caves and frozen moisture of the subarctic.

Though far less common than marine and terrestrial species, the 500 or so freshwater species are mostly disproportionately long-legged asellotids that barely resemble terrestrials. There are also several peculiar, eyeless, freshwater species in Australia and New Zealand from the suborder Phreatoicidea. Although they do not look so incredibly different, few people would recognize any of the freshwater aquatics as roly poly relatives.

<u>Body Parts</u>

The isopod body is divided into three regions: head, thorax, and abdomen. The head of the isopod is not attached to a large carapace as with many crustaceans or most arachnids. The large thorax also looks quite different from insects because the upper surface in most species is divided into seven equally large plates each with a pair of legs. Also, the thoracic plates seem to flow into the plates of the abdomen so they are often confused with the millipedes that do not have a clearly divided thorax and abdomen. As with centipedes and most crustaceans, there is an extra pair of mouthparts called maxillipeds because they are adapted from the front pair of legs. However, unlike lobsters or centipedes where maxillipeds come from a segment behind the head and look like adapted legs, isopod maxillipeds do not betray their origin. In most isopods this means the head is imperceptibly fused to the first (<u>not</u> the first visible) thoracic segment. Origination distinctions may be why the next seven segments of the thorax are together termed the pereon. Each of these large segments has a pair of legs called pereopods.

The abdomen of crustaceans is known as the pleon. The upper segments are called the pleurites while the appendages on the pleon are known as pleopods. The terminal, tail-like, often triangular segment is known as the

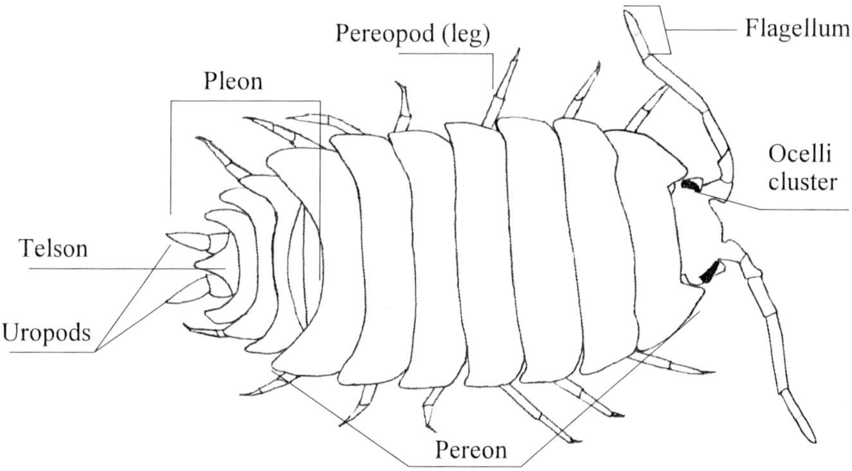

telson. As with many other crustaceans, the male pleopods are secondary sexual characters but unlike familiar crabs and lobsters, the female's are not adapted for holding eggs.

The pleopods can be used in swimming but the primary use is for breathing. Other familiar crustaceans have gills on the thorax (lobster, crab, etc.) or on the legs (brine shrimp, triops) so the abdominal pleopods can be adapted for holding eggs or swimming. Aquatic isopods normally pulsate the pleopod gills to keep water flowing across. Above the water these gills can work if kept moist, while littoral species wet them with water from time to time. Most of the familiar terrestrial species have "lungs" on the pleopods. These show up as white markings or squares that can be helpful in certain identifications. Whether they have lungs or not, terrestrial species are resistant to drowning and may walk around for hours submerged but they will eventually drown.

Crustacean eyes are far more variable than those of other arthropods. Isopods run the gamut. There are eyeless species, but in captivity they run from the single or double pair of ocelli found in the micropods, to eye patches of most large terrestrials, to the spectacularly complex compound structures seen on our shore-dwelling *Ligia* species. Some crustaceans have impressive compound eyes containing up to sixteen different types of color receptor cells (compare to humans with three and butterflies at five). *Bathynomus giganteus* eyes have nearly 4,000 facets (Caldwell 2008).

Unlike insects with two, crustaceans normally have four antennae. This is true for isopods but the inner pair for most terrestrials is so reduced they are difficult to locate without a microscope. The last segment of the antenna is called the flagellum. It is usually composed of two or three articles for common terrestrials but can be comprised of a dozen or more bead-like articles in shore-dwelling *Ligia* spp., but also members of the families Trichoniscidae, Mesoniscidae, Styloniscidae, etc. The flagellum often ends in`terminal setae which can help in identifying a species.

One feature that sets crustaceans apart from insects is the presence of branched appendages. This is not very noticeable in familiar isopods since the branched pleopods are small and closed but the branched uropods of sea slaters and *Asellus* are notable. Uropods are the terminal abdominal segments on either side of the telson that make up the swimming tail of crayfish, shrimp, etc. and swimming isopods. In most terrestrial species the uropods are long and thin and may have a sensory function somewhat like the cerci of insects. Whether an appendage is branched or unbranched may be seen labeled as uniramous (uni = "one", and ramous = "branch") and biramous (bi = "two").

Taxonomy & Systematics

Isopod taxonomy can be easy or difficult to follow depending on knowledge of specific morphological characters and the number of species. In northern areas the lack of variety usually makes identification by a few key features possible but if you do not know these features you could spend hours with a quality stereo microscope and discover nothing. Isopod taxonomists concentrate on certain groups or regions and finding one who could identify a small Central American species such as the "jungle micropods" even to the generic level with certainty is beyond difficult.

In the case of our three most common volvating species, identification is very easy. *Armadillidium nasatum* has a large, upturned plate between the eyes (in other animal species with a nose-like process the name is spelled nasutum), *A. vulgare* has no such process and *Cylisticus convexus* has spiky uropods that look nothing like the flattened plates of the previous two (*Armadillidium* tails resemble a lobster's). There are a number of other differences, but these alone make identification very easy. In Europe there are a number of additional *Armadillidium* but most are similar in size and color to variation common within the immature color range of *A. vulgare*. Other, less common volvating species are found in the genus *Venezillo*. There are eight US species of which only one, *V. parvus*, is introduced (Jass & Klausmeier 2001). They are easy to differentiate by locale because they are not common everywhere but are most easily separated from other genera because the telson ends in a wide blade rather than a flat point. There are also a few native *Armadillo*, *Armadilloniscus*, and *Tylos* species but native species are rarely encountered. The *Armadillidium* are members of the family Armadillidiidae, while *Armadillo, Cubaris,* and *Venezillo* species are in the Family Armadillidae, *Armadilloniscus* from the Family Scyphacidae, and *C. convexus* is from the Family Cylisticidae all from the infraorder Ligiamorpha, while *Tylos* are from an entirely different infraorder, Tylomorpha (Jass & Klausmeier 2001).

When it comes to representatives of the family Porcellionidae, specifically *Porcellio, P. laevis* has a smooth carapace, while *P. spinicornis* has a lumpy carapace like *P. scaber* but its color is mottled with yellow patches while the head (bearing a prominent middle lobe) always remains black. Also, there is a small spine projecting forward on the third antennal segment and the face is indented. Most of the "*P. scaber*" in my yard are *Trachelipus rathkii* (Family Trachelipodidae) because though they have the same looking head, antennae, and overall body features if you flip them over and look at the pleopods there are five small white patches rather than two larger ones (these are the lungs). *Porcellio dilatatus* are easy to differentiate from all the above because they have a rounded rather than pointed telson. The four native USA species of *Porcellio* cannot even be found in photographs. *Oniscus asellus* (Family Oniscidae) again looks similar to *Porcellio* and remarkably similar to some color forms of *P. spinicornis* but is easy to pick out because there are no white areas on the pleopods, and their flagella has three articles while *Porcellio* and *Trachelipus* species have two. *Porcellionides* species tend to be a little smaller and the abdomen is notably narrow compared to the body. Their body color is usually a spotless powdery gray. *Philoscia muscorum* (Family Philosciidae) are difficult to confuse with all the above since they have smooth surface and unique spotting pattern in addition to a tapered abdomen, but they look similar to other members of the family Philosciidae including *Atlantoscia* and *Floridoscia* from Florida and even the various *Littorophiloscia* from coastal areas. Philosciidae can also be confused with small sea slaters in the genus *Ligidium,* but sea slater eyes are more complex and the last antennal segment is broken up into a number of small articles, very unlike the three of *P. muscorum*.

When it comes to very tiny species the white micropods were long the only ones traded with even a genus name, because *Trichorhina tomentosa* is a widespread, tiny, white species, not because it was originally keyed out by a taxonomist. Costa Rican jungle micropods were named eleven years after they entered the trade. There are some tiny, temperate *Trichoniscus* species found mostly in the eastern states that may be confused only in that they also reach only a few millimeters at adulthood. However, they are mostly smooth, are often reddish in color and the "eye" is comprised of three ocelli. Another adventive, *Trichoniscoides sarsi*, looks similar to the *Trichoniscus* spp. but each eye is just one ocellus like *T. tomentosa*. Some tiny native species are similarly small, but quite rare and may be found in certain ant nests.

Isopods in Disguise

Cryptic coloration and hiding under logs and stones help isopods avoid being eaten. They have no chemical defenses to poison a predator or impart an unpleasant odor or taste. Running and dropping to conceal themselves in the closest crack and tightly gripping to surfaces are part of the defense. Some species pull in the legs and adopt a defensive curl. When in this curl they may refuse to move for many minutes in hopes the predator loses interest. It is amazing how many small predators lose interest in something as soon as it stops moving. However, the most conspicuous and fascinating defense is the ability of various unrelated species to transform into spheres.

The transformation from a natural creature with legs and antennae, to an unnatural globe or sphere most likely has the least to do with disguise but plays a pivotal role in escape and evasion of predators. A round ball will be much harder to see among small rocks and gravel and many volvating species live in rocky areas. Round shape seems unlikely as camouflage since many do not stay in the curled position long enough for a predator to get very far. Cultured specimens often refuse to stay curled up for more than a second. When a predator tries to grab a ball it can be much more difficult to grasp. The transformation between the two states can cause a predator to lose its grip. A ball will bounce, roll down slopes, and fall into holes. A ball can also be difficult to carry or bite.

Conglobation and volvation are alternate terms used to describe the act of rolling into a ball. In this text volvation will be used because it is more precise and less cumbersome. Conglob- means together or with globe while volv- means to roll or roll up. Conglobation is often used for cuckoo wasps which roll into a glob but do not form an appendage-less perfect sphere while volvation is common usage for pill millipedes which are much more similar in form and function to pillbugs.

Some volvating species are found in marine and littoral environments, but a transition in morphology and behavior is seen from sea to shore and further towards the deserts. Rolling into a ball is one strategy to survive "difficult" environments, specifically arid climates. *Venezillo* spp. range widely over North American deserts. The genus *Armadillidium* is the richest in the Mediterranean, often found in dry, rocky mountain areas.

Isopods are probably the first volvating creatures on earth since they are an ancient marine group whose body readily adapts to this form. Other volvating creatures include mammals like the armadillo and pangolin and a large assortment of unrelated invertebrates. In millipedes, three different orders have members that display this defensive behavior (keep in mind all the isopods are in a single order). Millipedes are primarily terrestrial and the oldest almost certainly rolled into a spiral, but more recent types can roll into a sphere. A number of cockroaches form hemispheres somewhat like the isopod defensive curl but only a few rare species such as the tropical Asian Pillbug Cockroach, *Perisphaerus pygmaeus*, are able to fully roll up.

Although the result may seem the same, the way in which the body plates fold up together can be quite different among arthropods. In pill millipedes each body segment has a recessed lip and tapers towards each side so the plates come tightly together at the central sides of the ball. In isopods the sides are less tapered and fold under the segment in front of each. Pill cockroaches have similar thoracic segments but three versus seven and the relatively large abdomen does not have segments that extend lower than the body, so they do not overlap when the sphere is formed. A close look at the edges can reveal what animal is being depicted but there is a much easier way to differentiate isopods from other volvating arthropods. Eight small, dorsal pieces that make up the abdominal tail are visible. In millipedes and cockroaches, the last segment is a single large plate. In volvating beetles the entire sphere is only three segments.

Despite incredible differences, the confusion of isopods with pill millipedes likely has resulted from lack of familiarity with either. While pill millipedes have 40 or more legs versus the isopod's 14, they can stay coiled for hours, so it is impossible to count legs on a healthy pill millipede. It can be difficult to get isopods to stay in a sphere long enough to focus and snap a photo. Increased interest in isopods meant confusion was becoming less common in 2006, but then US imports of pill millipedes were stopped along with any confusion. Years afterwards someone would come along now and then asking where to find the "giant green pillbugs" which are actually Madagascan millipedes.

Common Names

Terrestrial isopods are known by many names which may or may not have to do with their biology. The most common are pillbug, roly poly, slater, sowbug, woodlice, and "potato bug". Pillbug is a great, lasting name but it only refers to volvating species. Roly poly is also widely used for volvating species but usually only by very young children. Slater is normally reserved for the large, long-legged shore species, but slater is used for all the common garden varieties in New Zealand (Forster & Forster 1980). The name sowbug means "female pig"-bug and seems to have fallen out of favor in common use. Usually if an animal does not roll into a pill it is just an isopod. The name woodlice refer to their common appearance on wood, but lice are parasitic insects and isopods are not found on live wood, so the reference is based on ignorance, not a unique understanding of marine biota. This term is rarely used in singular because the reference is to a group. In the northeastern US where I grew up, they are often called "potato bugs," but potatoes are not a crop grown commercially in my area due to soil conditions. Isopods do not burrow deeply nor have a special affinity for buried potatoes. In the western US, "potato bugs" are massive, burrowing crickets.

Some common names are more correct than others. To begin, there are pillbugs, and there may be pill-bugs but there are no pill bugs. The space denotes systematic accuracy of the noun (not the descriptive verb) and isopods are not bugs. Consider: a ladybug is not a bug, nor a ladybird a bird, but a lady beetle is a beetle. This rule would apply to woodlice, sowbug, and potatobug if I were planning to use those unhelpful terms elsewhere in the text. The hyphen is considered unnecessary in codified naming schemes but could be used in place of a contraction to display systematic inaccuracy.

The common name pillbug is generally straightforward but still has been debated. The name is like a snail to a slug (a slug with a shell is a snail). Just because many marine snails are more closely related to marine slugs and certain freshwater snails while terrestrial slugs are more closely related many freshwater snails, this does not mean that systematics can be used to make shell-less slugs snails. Some people have tried to award taxonomic significance to pillbug and claim that many "pillbugs" cannot volvate while many "non-pillbugs" can. However, the common name refers to a physical attribute and only those isopod species that roll up into a sphere or "pill" can be called pillbugs.

The name isopod is somewhat of a misnomer when reduced to its root meaning since it comes from the words "same" and "foot". This refers to the second pair of thoracic legs ending in single claws, the "same" as posterior pairs, when this pair terminates in pincers for most other crustaceans. However, the first pair of thoracic legs is modified into tiny mouthparts so different they must not have been recognized when the order was first named.

Chapter 2. Biology

Terrestrial isopods are helpful creatures that feed on decomposing plant and animal matter and help to recycle these materials back into the environment. They are considered beneficial as they break down dead plant material and provide food to lizards and birds (Werner & Floyd 1994). Their beneficial biology and lack of any real commercial interest was the reason interstate transport was not previously considered for regulation. The lack of regulatory consideration/requirement was according to a presentation given on permitting requirements at the 2009 American Tarantula Society Conference in Arizona by W. Wehling, head of USDA permitting, but regulatory structure is being promulgated as of 2018. There are 113 types of terrestrial isopods in North America, north of Mexico with only 2/3 of these being natives. The fact agriculture has accidentally introduced and spread 1/3 of present US species, mostly from Europe, around the US may play a role since there is intrinsic difficulty in regulating accidents, introductions already occurred, and none have proven harmful. Still, if a keeper must get rid of a culture and it cannot be traded or given away, it is important to remember captive stocks should never be released and old substrate should never be discarded outdoors unless it has been thoroughly cooked. In many areas it is illegal to release non-native fauna of any type even if it can be found elsewhere in the state. Depending on local rules, a toilet may be an option for disposal of the tiny animals since even aquatic isopods would not survive minutes if flushed (any imaginary sewer survivors would then be eaten by the albino alligators patrolling the sewers).

One reason terrestrial isopods were not previously kept as pets was small size. There was no widespread technology to photograph such small creatures clearly or with reasonable time and expense. Today anyone can take and a share a great photo for free.

Biphasic Molting

Biphasic just means there are two stages. Members of the Order Isopoda undergo a strange sort of molt where only the front half of the body molts and the back half is shed later after the exoskeleton has hardened (the rear portion can molt first but the order is immaterial to function). The front half includes the first four pairs of thoracic legs while the back half contains three pairs of thoracic legs. This affords them mobility during molting since they can still run around with only six or eight legs operational. In terrestrials, the front eight and back six are virtually indistinguishable but in many aquatics like *Asellus*, the two leg groups are different in size and shape.

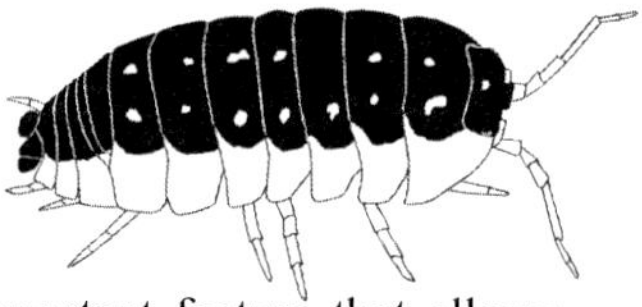

Biphasic molting may be the single most important feature that allows isopods to master the shoreline so fully, especially in rocky areas where burrowing is not possible for other creatures. Even terrestrial isopods may live in areas of periodic flooding during which a long period of restricted mobility would be deadly. Many other crustaceans like crayfish and crabs have a large carapace or cephalothorax covering the body so there would be no benefit to molting the back half at a later point in time. Insects do not have enough legs (split 2 and 4, two legs would offer little or no mobility) and use cell construction or hanging for protection. Myriapods have plenty of legs but dig underground burrows for molting safety. Arachnids may have enough legs but employ hanging or burrow construction to reduce the chances of being discovered by hungry animals during this vulnerable time. Lack of biphasic molting may be why none of these other arthropods have so thoroughly conquered such volatile habitats as the isopods nor crossed over notably between land and sea.

Other than the unusual biphasic aspect, molting is like other familiar arthropods. A new, softer exoskeleton develops below the old exoskeleton which is partly dissolved at the same time from underneath. As usual, this process results in a milky or gray appearance to the exoskeleton, sometimes a few days beforehand. At first the new exoskeleton is soft, and walking is not possible. Afterwards the exoskeleton hardens through chemical processes which take many minutes to hours for the new exoskeleton to obtain usable rigidity and up to a few days for maximum hardness. The process is known as sclerotization because it most commonly involves the protein sclerotin, but other proteins and calcium salts can be used in the hardening process. The remains of the exoskeleton are often consumed as soon as the molt is complete though other tank mates may eat it before the molting animal has the chance.

Social Behavior

As a group the isopods range in interaction from isolated individuals to rigidly defined and sophisticated social societies. Parasitic marine species usually do not see another isopod unless they are permanently attached to a mate or just happen to be on the same host. Many of the marine species are gregarious and found in groups like nearly all those on land and in freshwater. All live together in loosely defined groups and interact using the antennae. Most species exhibit limited or no concern for the composition of the groups. It is not unusual to find more than one species of isopod in the same aggregation on a sponge in the sea or under a log in a nearby woods. Often members of the same species congregate which could relate to mating, but aggregations can consist of all females or immatures. Populations of related species do seem to aggregate under a particular rock or log (and migrate together to new shelters) but these might be explained by dampness preference and shifts in the moisture level caused by external changes in the environment. Some of the large *Ligia* species found on rocks on the shore display an advanced hierarchy of individuals based mostly on size. This seems apparent when they are observed but they do not cooperate in any endeavors.

Active cooperation in terrestrial species seems to be related almost entirely to moisture conservation and can range from almost incidental to the most complex behaviors observed so far. Several large, non-volvating species display a behavior that can be termed as bunching (Kneidel 1994). Under dry conditions they actively cluster together and overlap bodies. This same clustering behavior is also known to offer an effective defense in some cockroaches from seasonally dry areas, notably the Madagascan Hissing *Gromphadorhina portentosa*. Bunching can be an effective defense against desiccation. However, which animals get the outside position and the immatures increased protection from fitting underneath are incidental. Related animals likely live nearby in nature but relation and species does not limit the behavior. The most impressive social defense against desiccation is seen in *Hemilepistus* species found in Old World deserts. Some of these have been studied for their rigid social roles created by the need to build and defend burrows in otherwise inhospitable desert areas. The burrow can only be constructed in the spring when the ground is soft enough to work. A monogamous male/female pair work together to build and defend their burrow with the help of their progeny later (Linsenmair 2007).

Unlike grasshoppers and several other non-predatory invertebrates that live in loose groups but cannibalize compromised individuals opportunistically during molts, this is density controlled in isopods. They do not bother each other unless severely crowded and this can be mediated by adding shelters. Still, a damaged and dying or dead animal can be rapidly consumed. Starvation usually does not change behavior unless it is combined with overcrowding. Swarming behavior is less common but may be seen when overcrowded and well-fed. When crowded they exhibit swarming behavior and rush in at the first smell of hemolymph. Nevertheless, even when a culture is both starving and severely overcrowded, most individuals molt successfully without being consumed. Even if an enclosure is busting at the seams with starving isopods, they usually do not swarm on exposed, molting cockroaches, millipedes, etc. unless those animals are compromised (a broken bleeding section present due to unrelated molting difficulties).

Right, Left, Right...

I watched a Japanese game show once where contestants bet on the validity of "myths". The only one I remember from the show was about isopods alternating directions each time they run into a barrier. Whether they start left or right does not matter but the next turn is always the opposite direction. It sounded far-fetched and I do not know if anyone won. However, when that little isopod ran the maze it went right, left, right, left like clockwork.

Gender Differences

Adult males and females of many terrestrial isopods do not look much different. Color differences between genders exist on a few species though they tend to be subtle or inconsistent. *Cubaris* sp. "Jupiter" males and females are a different shade of yellow. Male *Porcellio silvestrii* are orange and females are gray, but not always. *Porcellio werneri* males have a wider white colored margin, flatter body, and more pointed pleonites. Many of the large Iberian *Porcellio* (*P. bolivari, P. expansus, P. magnificus*, etc.) are easy to sex because the males have extremely long uropods and larger antennae, but most terrestrials are tough to tell apart. A few widespread species like *P. laevis* can exhibit enlarged uropods on adult males but the difference is minor. Young adults will be breeding for a year or two in captivity before they are large enough to be differentiated because the long uropods only develop notably with old age. Pleopods at the base of the abdomen are male secondary sexual organs that show adaptation in shape. These male appendages can be the only structure useful to differentiate closely related species. Also, in some species the rear pereopods/legs can have extra spines or hooks the female does not possess. Differentiating by pleopods or small spines on the pereopods requires a loupe or stereomicroscope and possibly dissection of the animal.

Another near-microscopic way to separate males and females of most species is by checking for the presence of oostegites between the legs. Even females that are not currently reproducing will have small bumps near the inner sides of the walking legs from where they are formed. For most terrestrial species it is difficult to tell without unrolling or spreading out the animal when the female is not holding eggs or young. If she is holding, very little pressure will pop the oostegites and release eggs or young. Eggs released will just die. Another problem with sexing is the keeper can easily damage or kill specimens trying to pick them up, let alone forcibly unrolling body parts for gender determination. If a specimen is near a molt, it will probably be killed before the condition is noticed. When there is such a small number that sexing is desired, it is not worth killing a single male or female (whatever sex there is plenty of will never be the one that accidentally gets killed). Once a keeper becomes familiar, sexing will become second nature and chances of damaging a specimen greatly diminish and yet gender determination is purely academic since there is usually no benefit to risking the death of anything in short supply. Males and females are all kept together safely.

Remember that for many species, females greatly outnumber the males. A disproportionately high number of females is normal for wild-caught animals whether attributed to microbes, differences in natural male survival, or gender biased trapping methods (Scott et al. 2001). A notable female bias extends to some captive terrestrial stocks but sometimes the bias is reversed and might relate to gender switching.

Reproduction

Fertilization is direct as males transfer a spermatophore to the female. Males usually sit directly on top of the female, not riding off the back like turtles or most beetles. Mating is a rarely observed event for most terrestrial species, but large male *Porcellio laevis* and some *Armadillidium,* especially *A. klugii* and *A. maculatum,* are commonly seen guarding the females. A few species reproduce through parthenogenesis, that is there are no males and each female forms eggs that develop without fertilization. This is how *Trichorhina tomentosa* normally reproduce. Other species can have a large proportion of females to males, but this does not mean parthenogenesis is occurring.

Mature females produce several eggs deposited inside the marsupium between the legs. The marsupium is formed by plates called oostegites that protect the eggs and allow the females to carry them around. The mother carries them everywhere she goes and keeps them safe. Oostegites look like one membranous sac but are multiple, hardened, overlapping plates that are easily popped apart by minimal pressure. If a female is grabbed, the eggs or mancae may start oozing out. Eggs released prematurely will not be put back by the female and they will die.

The number of eggs is determined by the species and the size of the female. Larger females tend to produce a greater number of eggs, sometimes more than double that of a small female, rather than larger eggs. An older, larger female can carry as many as twice the eggs of a small female of the same species. However, the range is determined by species as *Trichorhina bioccelata* often produce less than six while *Armadillidium nasatum* can produce close to a hundred (half or twice either number is not going to bring them close together). Still the lower volume producers can be more productive because they reproduce continually, start out larger, and mature faster. One of the larger species and one with large broods, *A. vulgare*, can produce young less than once a year even when conditions are good.

Eggs hatch after three weeks to three months according to species and temperature. The hatchling or first instar is called a manca. Mancae often molt to the next stage within minute of hatching. Manca is only the name for 1st instar pericaridians since this stage has six rather than seven thoracic segments. This is difficult to see even under magnification since mancae are tiny and pale. Fortunately, this type of naming is rare or the crustacean worker would have to memorize thousands of names to cover every minor morphological difference as there could be two dozen names to remember for one tadpole shrimp (*Triops* spp). Unlike most marine crustaceans that have strange-looking planktonic larvae, the 1st instar marine, freshwater, and terrestrial isopods normally look very much like a miniature adult except they are pale in color. They will stay pale for the first two or more instars and may not obtain full coloration until maturity. Some species like *A. vulgare* and *L. pallasii* display unique juvenile coloration that is not a simple transition between pale young and adult patterns.

<u>Life Cycle</u>

Isopods are slow growing compared to feeder insects such as fruit flies that take about ten days from egg to adult or crickets that require two months. Some of the fastest like "jungle micropods," take nearly three months while *Porcellio* species must develop for six to nine months. An *Armadillidium* female can take more than a year and a half before she is ready to produce her first brood. Unlike fruit flies and crickets, that die a few weeks or so after laying eggs, sexually mature isopods continue for years and continue to reproduce and grow. Differences in environmental conditions such as food and temperature, as well as differences between species and even individuals, can have a huge impact on the time spent in adulthood. However, most terrestrial species live about two years after reaching maturity. The freshwater species rarely live a year while the incredibly variable marine species probably range in adult longevity from weeks to decades.

Chapter 3. Clean-up Crews for Bio Active Habitats

The history of keeping isopods is the history of the clean-up crew. Isopods started to gain some popularity in the early 2000s almost entirely for their use in keeping the habitat for other pets healthy. The oldest text reference with an explanation and the term "clean-up crew" is probably the reference in *Giant Centipedes the Enthusiast's Handbook* (2003) though it goes at least as far back as the late 1990s when I first began trading the isolated "Spanish orange isopods" and "big white springtails." The idea came from the clean-up crews commonly offered for the mini-reef aquarium. The concept is similar though isopods are mostly used for control of rotting, excess scattered food and prey materials that otherwise result in contaminated air and harmful mite and pest infestations. Marine clean-up crews are composed of blue-legged hermit crabs that eat hair algae, and turbo snails which feed on slime algae, though there are the less commonly used emerald crabs that target *Valonia* spp. algae, and peppermint shrimp that eat *Aiptasia* anemones. Terrestrial isopods are superior to reef clean-up crews since they can be easily reared in captivity, can be shared among enclosures, and rarely need to be replaced (turbo snails, blue legged hermits, etc. are not realistically reared in captivity and must be purchased at regular intervals as they die). Moreso, isopods can be employed as feeders, as a clean-up crew or a combination of the two. Like reef clean-up crews, they add some life to the vivarium and help establish balance in small artificial ecosystems.

Uses are varied. Most isopod species are employed to keep damp terraria clean of rotting prey and prey parts, certain fungi, and rapidly decomposing plant matter. Low humidity is not a reasonable alternative to control pests for many creatures since it kills the main inhabitant as well. Isopods are most often put to work in gecko and amphibian terraria, whipspider and humid tarantula cages, large forest scorpion enclosures, cockroach containers, and centipede jars. The items isopods feed on can poison the air in the cage or lead to infestations of grain mites, phorid flies, fruit flies, or other destructive and annoying cage pests. They eat rotting items and prey pieces the main inhabitants either are too large to find or are too picky to eat. Many large predators discard prey parts or a large pellet of incompletely cleaned chitin. Isopods prevent certain types of entomophagus fungus by removing its source of food. Isopods also consume the decomposing fruiting bodies of mushrooms but do not feed on most live wood or leaf fungi. Fortunately, the fungi they do not eat is harmless to other invertebrates. Most predators will ignore isopods and allow them to do their work unless they are not sized correctly, or the keeper forgets to feed the predator for a while and it grows very hungry. Large centipedes and small tarantulas are especially prone to taking out every janitor if better suited prey is not regularly offered.

Isopod clean-up crews are incredibly useful and helpful but they are not magical. They simply outcompete enclosure pests and if the population is too low, they will not work efficiently. For a clean-up crew to work it should be chosen for the appropriate size and there must be enough to consume all rotting materials. If too many are added the excess will starve. The exact number will vary greatly by how much food is being eaten, how much is being fed, various aspects of the enclosure, and the type of isopod. Fast breeders can build up to good numbers after a few months and few may be needed. If they also serve as food for the cage inhabitants, the culture container should be separate.

It is important to choose the right janitor. Some types will be poorly suited or quickly eaten. Micropods are often the best clean-up crews because they are least likely to be eaten and they reproduce rapidly. Micropods do well in hot, wet terraria but can be impeded by tiny dry barriers or can reproduce so much they need to be manually thinned out a few times a year. Striped dwarfs and medium sized species like *Porcellionides* are a sort of halfway point between the micropods and the big species. The Orange *Porcellio* are commonly used because they are useful cleaners, are easy to control, survive a greater range of environments, and provide food and beauty in the bio active habitat. Cryptically colored and species more prone to burrowing are employed when the singular goal is an invisible clean-up crew. Pillbugs and the larger *Porcellio* are generally suited to less damp and cooler terraria and seldom outgrow their welcome. Chapter six lists stocks that are commonly available and addresses individual considerations.

There is a Dark Side

Isopods should not be used with beetle larvae, millipedes, or anything that directly competes with them for food. They might even outcompete less virile, rare, roach species, but since the latter are finicky it would be tough to prove a negative outcome with hard data. In small numbers isopods are harmless to everything except exposed beetle pupae, but when they number in the hundreds and thousands they can eat soft items that do not move. They usually will not bother millipede eggs in sealed capsules but any cracked or broken capsules will be emptied by hungry maws. Molting animals can also be at risk but problems are rare because the movement of a molting arthropod will scare off the isopods and many hang out of reach of the isopods or have protective mats of hair. Reports of isopods attacking molting animals are not common, and probably the result of a sick animal dying during the molt and then being consumed. If factual, such problems only occur when the isopod population is ridiculously high in the cage. Isopods are not predators, but they do take advantage of nonmotile available food. If they did not readily feed on dead invertebrates, they would be worthless as a clean-up crew.

Large numbers of starving isopods will exhibit swarming behavior. Hungry animals are attracted to feeding frenzies and they do smell hemolymph (invertebrate blood). My whipspider cages remind me of bat caves because as soon as a ball of chewed chitin is dropped the isopods swarm and completely cover and tear it apart. This also happens to a half-eaten cricket. If a whipspider were to fall while molting it seems it might suffer the same fate though I've not yet known one to disappear this way and if it did fall it would die anyway. Its imminent death would only be 24-36 hours more rapid, not more certain, and reduce time suffering.

Compare and Contrast: Springtails

A discussion of clean-up crews would come up short if springtails were not considered. The singular complement or competition for isopods is springtails and presently only a single species, the "big white" (also sold as tropical pink and other names) has proven of any real function. It is easy to raise all sorts of springtails but the vast majority are just too small and not prolific enough to function as a rapid clean-up crew. It is not an either or proposition, isopods and springtails are often used together in bio active habitats.

Springtails are useful in damp enclosures with millipedes and beetles because they cannot harm or unearth vulnerable stages. They may also be needed if

Big white springtail

the cage inhabitant readily eats small isopods. The biggest drawback to springtails is the cage can look rather infested. The visitor almost always asks, "What are those things all over?" Isopods are relatively discreet and even huge numbers can remain unseen. Many of the isopods hide during the day even if a colorful form is chosen whereas none of the cultured springtails are cryptic in color and springtails climb and jump all over the cage. Despite springtail infestations that look like a documentary on introduced wildlife in Australia, they do not harm, stress, or eat commonly kept arthropods.

Versus Predatory Mites

Predatory mites are beneficial helpers in certain circumstances but in a terrarium where food is in supply to keep them going, a large culture of these mites is harmful to some creatures. I have found them functional for control of grain mite infestations such as for infested, damp rhinoceros beetle substrate, and dry substrate for grain beetles and dermestids. In the case of grain substrate, tossing out and replacing the media is not a real option since all ambient mites cannot be completely eradicated and new media will be rapidly re-infested. It usually takes three weeks to a number of months, but the more massive infestations of grain mites are usually the most rapidly controlled. Even hypopus stage grain mites are consumed. They often completely eradicate grain mites over time and then die out themselves. Predatory mites can eradicate nematodes and reduce fungus gnats, but it takes months and reduced or absent food. These would be superior to isopods in most applications where the singular goal is grain mite control except they do not work in many large or heavily populated cages. Predatory mites can work on various media infestations of different moisture levels but when the source of the problem is dead arthropod bodies or parts of any volume, they have very little impact on pest populations. Predatory mites can be expensive and are difficult to culture except in those enclosures where they are unwanted. Their presence severely depresses and can lead to eventual failure of springtail and Pillbug Cockroach cultures. Fortunately, they have no hypopus stage. Though large compared to grain mites, these are very tiny and do not negatively affect aesthetics.

Chapter 4. Culturing Isopods

Isopods are among the weakest and mightiest of arthropods. Individual specimens can seem weak since they are easily damaged by handling, and most can die rapidly when held in the hand (try carrying some home from a nearby wooded area in this way). They also rapidly die from exposure and dryness. I have seen large numbers of dead in nature, possibly due to freezing. Nevertheless, isopods reproduce and grow at such rates losses are insignificant. Common culture stocks are effortless to care for and keeping a culture for decades would require little expenditure of time or energy. This hardiness makes them an excellent choice for culturing. The following sections will outline experiences that should reduce the chances of accidentally killing isopods.

Culture Container

There are quite a few options for containers used to house an isopod culture. The most common are plastic tubs, plastic shoeboxes, and aquariums. Glass or plastic jars are fine but the height is wasted space. Cardboard, wood, or even masonite boxes would not be a good choice since they are readily scaled and warp from the required humidity. Uncle Milton's Roly-Poly Playground is hard to ignore as the only cage designed and sold uniquely to house isopods, but it is a tiny children's toy. Originally, I maintained cultures in the molded plastic small animal cages but the excessive ventilation made it necessary to water substrate regularly. A little negligence could wipe out a culture kept for many years. There is no reason to add an unnecessary layer of difficulty, so I now use cages with retricted ventilation. Whatever container is chosen, consider starting with two cages if it is a difficult to replace stock. Small differences can add up. If one culture dies out it may be easy to determine the cause through comparison.

Climbing is not usually a consideration since isopods have no adhesive pads that allow climbing found on many other familiar invertebrates. Also, the common behavior of hiding may conceal climbing tendencies. Isopods are light-weight and can climb up the silicone bead in the corner of an aquarium. Small individuals and species readily climb up barely visible films of dirt on smooth container walls. Even a thin layer of condensation can sometimes provide tiny animals a means to climb very tall, smooth surfaces.

Moisture, ventilation, exposure, airflow, and humidity are important, interrelated considerations for choosing the culture container and lid. Humidity and moisture are terms often used interchangeably but humidity is the amount of water vapor in the air while moisture refers to the amount of water or water droplets in the substrate and on the glass. High water vapor levels may reduce moisture loss, but animals cannot drink it. Airflow increases water loss even in the presence of high humidity. I have seen greenhouses with 100% humidity that will dry out an exposed isopod within minutes. Substrate materials or wood can shield animals from water loss due to lack of humidity or excessive airflow. The need to retain moisture (rather than avoid predators) is most likely the greatest reason terrestrial isopods hide under flat objects or burrow.

Moisture gradients caused by different temperatures on opposite sides of a container can cause dangerous conditions where some animals become stranded and dry out though the other side is dripping wet and humidity high. This is common on shelves facing a wall. Water vapor condenses on the cold side and over time the warm side can become bone dry and the cold side dripping wet even with severely restricted ventilation. The container can be rotated occasionally, or the substrate mixed by hand every few weeks. Nevertheless, many big Iberian *Porcellio* and some *Armadillidium* prefer to have moisture gradients.

With all medium to small species, almost no ventilation is required. A solid glass lid is placed on the cage or a plastic shoe box with no extra ventilation. A few thumbtack holes are made in the lid of small plastic containers. For larger species like *A. vulgare* and *P. laevis* it is best to drill half a dozen or more 1/16" (1.5 mm) holes in the lid of a plastic tub or bucket. The determining factor is not always size, *A. nasatum* can average the same size as

A. vulgare but are not sensitive to limited ventilation. The ambient humidity, airflow, temperature, number of animals, food, and moisture of the substrate are all considerations. If adjustment of the moisture is required more than once every few months there is too much ventilation. If moisture builds up and animals seem to be dying for no reason there may be too little. The amount of air in the container also relates to ventilation since more open space reduces the need for ventilation.

Substrate

The bottom of the cage should have substrate for moisture retention and traction since isopods cannot grab onto smooth surfaces. It would be very easy to dry out isopods in a container with no substrate even if there were no ventilation. I have used sand, gravel, potting soil, composted manure, beetle frass, coconut fiber, and peat with good results. Some substrates prove overly sticky and discolor or slow down the animals too much. Others can lead to excessive mold growth over time depending on the conditions present (humidity, moisture, food, and temperature). Specifics are why a substrate that works well for one keeper may offer problems to another. Mold may take many months to show up, but substrate should be changed if there is a lot of mold growth. Molds may not directly infect isopods, but they can become stuck in excessive fruiting bodies and filamentous growth. Sand and gravel are inert but generally used for shore species when very wet areas are needed with standing water (since they do not suck up or wick moisture). Other substrates hold moisture and work equally well. Usually substrate deeper than an inch (2.5 cm) is unnecessary.

Some researchers use the isopods' own pellets and a solid porous base which may prevent some substrate problems. A 4/5" (2 cm) layer of plaster of Paris can be applied to the bottom of the culture container. It is cheap and easy to apply. The plaster can be mixed with biochar or activated carbon to retard fungal growth. This material retains moisture and absorbs excess water. The surface is rough and chemically harmless to isopods. Once it hardens, leaf litter is placed on top. The isopods quickly create a layer of organic matter from pellets and leaf fragments into which they can burrow (pers. comm. Vilisics 2012).

Shelters

Wood and cardboard pieces are commonly used for shelter. Other materials are either too heavy, too smooth, or too easily chewed up. Heavy materials will smash and kill a good number of creatures. Many isopods like to grip the underside of rough, flat objects. They prefer to have the body between a narrow space or held tightly against a surface. Corrugated cardboard is not usually used since small animals will get into the folds and become very difficult to remove safely. In some cases, it can grow moldy. In the case of wooden shelters, if removal of animals may be desired in the future for feeding or transplant, it is best to avoid a lot of cracks and holes for hiding. If it is necessary to remove an animal from a hole in the bark the shelter may have to be broken apart and some isopods will be smashed and killed. Some types of bark have tapering ridges that offer maximum shelter and still allow the animals to be easily dislodged by smacking the opposite side with the palm of the hand. Most materials will be chewed on by the isopods and the shelter will become thinner and thinner over time. Soft rotten wood can be replaced regularly or avoided since it can be chewed into nothing rather quickly. Despite its softness, commercially available cork bark is not consumed quickly. Tree bark from various trees can be used. I have used pine and cedar in a few cages, and it is chewed up like hardwood.

Shelters influence burrowing behavior. Some species that don't normally burrow may try to burrow in the absence of shelters. Also, large substrate particles like aspen shavings or leaf particles can act as tiny shelters since they are bigger than the isopods. Some species that normally would not burrow in fine substrates may be found as deep as large air gaps and large particles are.

Temperature

A broad range of temperatures is generally accepted and even the tropical species can breed readily as low as 65°F (18° C). Of course, there is a bottom temperature for each species where reproduction ceases but this is below the bottom range of most heated structures. Temperate species stop breeding but will walk around, and some even eat, at 40°F (4° C). Temperatures far above room temperature (72-77°F or 22-25° C) are not necessary and can harm some species. Warm temperatures may improve reproduction, but Mocquard et al. (1980) showed high temperatures only marginally accelerate reproductive onset not its duration. Seasonal studies of isopod reproduction are not applicable to captive situations unless cyclical temperatures are used to induce seasonal reproductive periods for academic purposes or in hopes of possibly inducing a difficult to breed species.

Most of the temperate species be safely kept in the refrigerator but there is no reason for it. Some insects are refrigerated for the purpose of delaying development or prolonging useful life but it does not prove useful for isopods. Rapid shifts may be harmful and temperate species can experience high losses during natural winters (a number of dead animals are sometimes seen under logs).

Daylight

Lighting does not play an important role in captive care of most species. I use mostly semi-opaque rearing containers placed on shelves far from the closest light source. I have also kept some in a drawer. While light leaks in these containers, even within a closed drawer, may still afford them some sort of day night cycle, lighting seems to be of little importance. Isopods do not like light and this aversion to light is easily tested by giving them a choice. They will feed during the day on exposed surfaces in reduced light and when food is placed beneath a piece of bark but usually avoid daytime feeding in exposed, well-lit areas. Since isopods are most active at night it seems extended daylight would only reduce active periods and be likely to slow down development. Many of our native species are found only in certain caves or cave systems (Jass & Klausmeier 2001) where light is probably not a major consideration.

Bhella at al. (2006) state they extended the photo period to 16 hours a day to enhance reproduction though the study had nothing to do with testing photo period results and they did not refer to a source for this idea. They likely did this because previous research on *A. vulgare* showed that lengthening photoperiods were an impetus for reproduction and extended the reproductive period (Souty-Grosset et al. 1994). There are no seasonal variations required for commonly reared captive populations and no cycle is required even for *A.*

vulgare to reproduce in captivity. However, the need for day length cycles (possibly affected by temperature cycles) may be why captive reproduction of *P. muscorum* is difficult. As such, it is likely only applicable to someone very interested in that species or being paid to study daylight periods.

Food

The primary recommended food is brown, old leaves that have been on the ground at least a few months. I have used leaves from various hardwood trees (oak, maple, beech, birch, aspen, ash, elm, locust, redbud, sweetgum, walnut, persimmon, plum, pear, etc.) and to a far lesser extent shrubs (various *Viburnum* spp., lilac, raspberry, rose, etc.) and grasses. The Indian almond leaves sold for terrarium use will work but can be expensive. Some leaves are tougher and less readily eaten but lead to strong growth, such as oak, while others like maple are more easily chewed but nutritionally deficient so a mix of leaves works best. Harder leaves can be ground up to make feeding easier. Weathered old leaves are preferred since most fresh leaves will not be eaten until they begin to decompose. Also, excess nutrients in fresh leaves can feed grain mite problems. Fresh romaine lettuce is a readily eaten treat but I have not used it regularly nor tested how it compares or contrasts with various leaves. Rabbit or gerbil pellets can be used since they are made of grasses but they are made from fresh plants (decay is primarily arrested when kept dry) and can invite pests and mold.

Other common foods include fruit scraps, dry dog, cat or fish food, and various vegetables, especially carrots and potato slices. Usually rather inexpensive or free food is used since isopods are not picky. It is important not to offer too much dog food or fruits as even a large culture will only eat so much. Excessive rotting food can harm the animals and cause pest infestations. In the case of any foods that rapidly decompose, only what will be consumed in three or four days should be offered and only every other week. Pet foods that sit in the cage for weeks and become engulfed in mold (if invertebrate pests have not found them first) may eventually be eaten but even more slowly. Melon rinds work well but also have a short usable life. Carrots and potatoes often do not rot for months if cut into large pieces. Isopods may chew caverns into large potato pieces and use the potato as a shelter. Still keep an eye out as these can suddenly decompose, especially in hot and wet terrariums.

Algae, mostly hair algae from freshwater aquariums, is highly palatable though it is initially swarmed and later forgotten. This is likely because the isopods are attracted to the nutritious diatoms that grow on the surface and what's left is not of much value, rather than boredom with a certain food. Other freshwater algae can be used but not blue-green. Marine algae seems less useful, especially macroalgae like *Caulerpa* spp. Sheets of algae and prophase kelp can be purchased rather cheaply from Asian markets and can be used in place of more volatile foods for easy care. Reproduction is enhanced when kelp is offered (if leaves are not available) but some of the oldest adults in the culture will die a few days later. Isopods enjoy eating Java moss almost as much as crayfish do, but it may not provide any nutrients. Lichen, a combination of algae and fungus, is also a good food.

Rotten wood can be a useful food since it can be left in the cage indefinitely but will result in minimal growth and reproduction without other foods. Even shredded aspen bedding for vertebrates is of some value as food since it decomposes in the presence of moisture. It is not highly nutritious, but a handful can be added so there is something to chew on when they have not been fed in a while or are tired of the normal fares.

Since they have a measurable percentage of calcium in the exoskeleton, chalk, cuttlebone and egg shells are often suggested as part of the diet. Some isopods chew through chalk and cuttlebone at a good pace. Most readily eat the film off the shells when first introduced, but otherwise the shells are too hard. These calcium sources certainly do not hurt anything but may add zero value. Leaves, pet foods, etc. contain calcium so my isopods never saw supplemental calcium prior to 2013. Pieces thrown into large cultures for months show the same wear I would expect to see on an inedible object of the same hardness.

Most of the larger species only produce young once or twice a year in nature and can require certain foods as an impetus for reproduction (McMonigle 2004). Photo period and temperature cycles might work as an alternative impetus (there is no evidence they can be used in place of quality food if one considers the cyclical nature of leaves in temperate areas) or for species that are not commonly kept. Dwarfs and micropods breed year round and do not seem to care what they eat within reason. For most species the impetus is rotten leaves though fresh hair algae from a freshwater aquarium can work even better. If only fruits and dog food are provided, the adults grow and survive very well but can produce few to no offspring. Females not given a food impetus for breeding seem to end up diverting resources to their own growth because constantly breeding females can stay surprisingly small. This seems logical but males kept in the same conditions also stay small and they cannot be investing nearly as much in sperm production. Overcrowding is different from other creatures in that it does not lead to stunting. Large, overcrowded cultures often end up producing large, and even some exceptionally large specimens. This may or may not relate to food competition. Cohort splitting where some individuals choose to invest in growth rather than reproduction is also possible.

Culture Contamination

Other than the basic housing considerations outlined, it is important to consider how to avoid or deal with contamination by springtails, mites, worms, fungus gnats, and accidental introduction of other isopods. These can be issues but are usually not the problem they are for other types of invertebrate cultures. Springtails may be added on purpose but can be extremely difficult to extract if they are unwanted. It may be necessary to start up a new culture or decrease moisture and food for long periods. Mites come in various forms but the most problematic are the grain mites that are the primary reason for using isopods as clean-up crews in terrariums. These can also cause problems in terribly over-fed isopod cultures. It may be necessary to switch to only wood and well-rotted leaves because freshly fallen leaves contain nutrients that can lead to mite infestations. However, I have never seen the normal pest, grain mites make it to the harmful hypopus stage in an isopod culture under a wide range of circumstances. Nematodes and earthworms are often introduced with leaves and rotten wood if these items have not been thoroughly dried or cooked. Worms usually do not have much effect on the culture but can thoroughly overrun the substrate. Dark-winged fungus gnats are annoying little flies whose see-through maggots consume wet rotten leaves with gusto. The adults like to fly into human noses and eyes. Gnats can be prevented through airtight containers with fine screen vents or greatly mediated by not mixing leaves into the substrate and not offering too much leaf litter at once.

Possibly the most annoying culture pest is *Trichorhina tomentosa* because it is the only thing readily capable of overrunning other isopod cultures. If these are in a terrarium or container on a shelf or stand above another culture they will almost always show up. These can overgrow other species and are mostly annoying because they can be difficult to pick out from immatures of other species. You only have to miss one white micropod. Other pests can mostly be mediated through changes in feeding that give isopods the chance to outcompete them, but not these since they are isopods. Nevertheless, if the main isopod is a temperate species and the culture container is not too large, it can be placed in a refrigerator overnight for an easy, rapid solution.

Different species are sometimes purposefully kept together in culture. Although half a dozen different introduced species are sometimes found together under the same log, possibly together with a native species or two, there are also centipedes, spiders, and other predators that it is likely one would not want to house with isopods. It is possible to keep different species of isopods together for relatively long periods but if one species become overly numerous it will eventually wipe out the other.

Transferring Isopods

In order to transfer animals for janitorial services, it is important not to kill them. Handling can be a noisy proposition. If you listen closely you will hear the exoskeleton cracking and crunching under stress. Hand counting is easy for large, sturdy *Porcellio* and some of the pillbugs, but even these could be damaged if picked up with a finger and thumb against a solid surface. The easiest way to handle specimens is to place flat bark or pieces they can climb on in the container and then tap it over the new terrarium, the hand, or a shallow container.

The counting container should be no more than a few inches high and have sloping sides or rounded corners and not deep scratches, punctured, or drilled areas. Small burs in plastic will tear the isopods apart as they are pushed up the side. For counting, animals as small as a few millimeters can be safely pulled up the side with a finger towards a waiting thumb. If too much pressure is applied, crunching noises will be heard. Damaged isopods are capable of regenerating and healing as they molt but most often the damage is severe and results in rapid death.

Chapter 5. Unnatural Selection & Captive Stocks

Beyond freshwater isopods purchased from Carolina Biological in 1990, and *A. vulgare* that were passed off as pill millipedes in 1995, my first foray into isopod culture began in 1997 with an opportunity to acquire iridovirus-infected isopods from a friend who had acquired groups collected by a researcher in Spain. These sounded exciting as they were supposed to be beautiful. I had seen many "blue" invertebrates that weren't very blue. When these arrived, I was floored. They even made electric blue crayfish look dull. The originals included a dozen blue *Armadillidium*, but they died after a few months and did not reproduce. The remaining animals were a *Porcellio* species.

There was a dozen blue *Porcellio* sp., four dozen solid gray, and a single orange animal. It took a few years and some selection before there were enough orange specimens to isolate a pure colony since no offspring appeared orange until the second generation. I had attempted to isolate blue specimens a number of times, but they always died after a few months. A few rare females would produce offspring before dying, but the resulting young were never born blue. Blue specimens would appear following a molt and did not change over until reaching maturity. Some would change over at initial maturity when they were not very large, but most changed during late maturity at full size. Unfortunately, in isopods the iridovirus that causes the blue color appears to b e 100% lethal to the individual within a few months, whether the animal is separated from the main culture or not. The culture was split into a few different cage styles and the blue remained a common sight (1-5% of adults) in the very damp cages but grew less and less common in enclosures that were kept less damp. The main culture (versus the orange isolation) threw blue specimens with decreasing frequency from 1997 to 2008 when the very last one was seen. In 2006 a single orange specimen from the main, mixed culture changed to a handsome purple. I had previously believed the orange were immune since none had changed over in the last nine years and never within the pure-breeding orange culture. The mixed culture threw the rare, variegated specimen which was encouraged but was far more rare than the orange (in the middle years 5-10% would be orange, 90+% gray, and less than 1% variegated. Through removal of countless gray and most orange a nearly pure-breeding, variably colored, variegated ("calico") stock was finally realized in late 2011 and perfected in 2012. It has been used since by various breeders in attempts to isolate yellows and other forms.

The iridovirus can infect *Armadillidium*, *Porcellio*, and probably most genera of terrestrial isopod. It is not easy to maintain in culture because it causes the death of the host unlike the iridovirus that infects crayfish (McMonigle 2004). Also, only a small number of adults are normally infected. The high proportion of blue animals in my original group was because they were collected from a larger wild population. The *Porcellio* would have been lost quickly like the *Armadillidium* had only infected adults likewise been acquired. No visibly infected specimen ever lived more than a few months. Keep in mind that infected animals will be differently colored or patterned according to the individual and species but animals with a tinge of blue or purple are not infected by this virus. A slightly purple *Armadillidium nasatum* cultivar (McMonigle 2013) and the naturally colored Costa Rican isopods can display a slight purple tinge, but they do not die and it has nothing to do with a virus.

At this point in time the major color morphs of any significance to the hobby all appear to be simple recessives. Initially these were the orange forms of *Porcellio sp.* and *A. nasatum*, and the pied forms of *Trachelipus rathkii* and *Porcellio scaber* but they have been seen or isolated in most captive species. The orange and pied form of *Armadillidium vulgare* may not be simple recessives while the white form of *Porcellio dilatatus* is probably similar but has not been isolated in culture. As simple recessives this means if the male and virgin female are both orange, pied, or white, 100% of offspring will follow the parents. A single stock can take years to isolate when, as in most cases, there is only one or a few founding adults. This is because the life cycle can take a year and the first offspring will be heterozygous for the recessive, that is they will all be the same phenotype (gray). If multiple unmated adults of the same recessive form are available, the isolation time can be reduced to nothing though it may still be a few years for decent production. I have found it is not a good idea to isolate as quickly as possible because the genetic bottleneck can lead to weakly performing stock. Also, the recessive gene can be accidentally weeded out of the parent group and result in complete loss if the small. isolated group fails or fails to produce both genders. If the grays are removed slowly as the recessive becomes more common, many of the resulting offspring are more likely to be het. When there are many or mostly het several recessives will be theirs rather than the one or two founding animals. Of course, the goal is to eventually remove all dominant and hets but doing this too early on can actually cause it to take longer rather than shorter time to produce a viable stock. A few species in recent years have been selected for both recessive genes so the dark spots are orange.

The calico and halfwhite combinations are far more difficult to pin down genetically. The calico seems to be a combination of a recessive orange and another more variable recessive gene because they show up only very rarely in gray and orange crosses. For years the culture would only produce a few of these compared to hundreds of oranges and thousands of grays. However, as the selected generations have progressed, the presence of resulting full orange offspring have grown more and more scarce and may have been eradicated. Removal of the dominant gray color genes and their disappearance from the stock over time makes perfect sense but if all the adults were really co-recessive around a quarter of the offspring ought to be fully orange. This seemed true early on when all three forms were present but after many generations full orange were not even 1%. Additionally, the calico color form includes a number of different patterns and selecting for a variety of expressions is easily done but difficult to explain genetically. I had originally been selecting for the gray splotches on an orange background, but the orange was often replaced with cream, tan, yellow, or a rare greenish so I switched to selecting for greater contrast rather than specific color. Initial pied forms of *Cylisticus convexus* and *A. corcyraeum* seemed to have been environmental not genetic as future generations under different conditions lost the trait. In the "calico" *Porcellio scaber* I isolated from 1997 Spanish stock over a dozen years, both sexes express the calico appearance. Bhella et al. (2006) encountered a different variegation in *Porcellio scaber* (stock collected in Canada) that was a sex limited trait transferred by males and observable only on females. Their study into the genetics of color lasted only a few years and involved no completely isolated lines, so some of the conclusions were probable rather than concrete. Conversely, there could be additional genes expressed in the bi-sexual calico.

When it comes down to the *Porcellio laevis* halfwhite form it would be far more difficult to plot phenotypes for the genetics involved because the same specimen can register as gray or white. *Porcellio laevis* produce a strange form called halfwhites, not because they are half white (though sometimes they are) but because a single specimen can be white one instar and gray the next. Grays can likewise change to white following a molt. It seems to be a recessive defect since it took a few generations to surface and after a number of selective generations nearly all specimens in the stock display this coloration, but each specimen is variable. In earlier generations, some of the affected animals would be partly or half gray and half white but the majority would be white one molt and gray the next. Presently, specimens develop a gray blush after unspecified molts, and most specimens now are yellowish rather than white. The halfwhite could possibly be the same gene that causes white individuals in other species, since Bhella et al. (2006) believe albinism in *Porcellio scaber* was not due to inability to produce pigment but inability to transport pigment precursors into some tissues. Though displayed differently according to species, it is likely most recessive genes for coloration such as those for orange, white, or pied are the same in different isopods (and possibly even crayfish). The halfwhite may simply be one way the partial albino gene affects the color of *P.*

laevis.

I tried to isolate a pale section *Porcellio* sp. (these might have been *Porcellio scaber,* but they were not as warty and grew as big as *Oniscus,* but were not *P. dilatatus*) for at least a decade. Unfortunately, after all reproductives were isolated for this trait only about one in a hundred offspring would follow. This certainly was not a recessive and so could not be isolated nor related to the gene causing the halfwhite form of *P. laevis*. After so many years producing a colony of maybe thirty, I grew frustrated and gave them away. I feel some regret since that keeper eventually killed them, but another two decades would probably make no difference as the prevalence never increased after isolating the reproductives.

Albino is a genetic flaw in natural color schemes caused by an abnormal failure to properly synthesize dark pigment. White micropods and other naturally pale species are not considered even partial albino because white is the natural state. The piebald *A. vulgare*, *Trachelipus* and *Porcellio* Dalmatians, and pale section *Porcellio* sp. are all piebald forms which are sometimes considered partial or incomplete albinos. Some of the white forms still have black eyes but a true or complete albino should have white eyes. Albinos can be expected to have a dark line for the heart. Even the superbly bright white form of *Porcellio dilatatus* has black eyes. Full albinos with white eyes have been isolated for *Porcellionides pruinosus* and *P. scaber*. It would be interesting to discover how the genes responsible for abnormal albinism in various species or pied forms are related to genes in more distant relatives like the abnormally white aquatic *Asellus* sp., other isopod species, and populations that are naturally white.

Morphological variations so far have played a little role in captive selection. The biggest reason is the primary character of interest, size variation, is almost impossible to isolate or select for. There is no ultimate molt and ultimate size is determined by external factors. One cannot simply isolate and breed the largest individuals. It takes years for sexually reproductive adults to reach full size and the eventual size is determined by food and habitat. Relatively tiny offspring will be producing young in the meantime so specimens would need to be isolated and labeled. Such an endeavor would require decades of close attention to detail just to see if it might be possible. The "spiny" form of *Porcellio laevis* is the only morphological variation I am aware of that has been selected for. The outer edges of each body segment pull up and back at the sides rather than merging into a rounded outline.

Pre-2013 Cultivars

Porcellio scaber

"Orange" If you go back to 2000, these were only traded as "Spanish Orange Isopods" because the original bloodline came from stock collected in Spain. These are a single source bloodline so if you have some you can be certain they are descendents of the original female no matter what name they are sold under. It is a rather hardy stock that does very well and older adults grow to 15 mm in body length. I have been trying to isolate the more solid coloration in which the orange color obscures the dark line of the dorsal abdominal artery but after twenty years and countless culls less than a third look this way. A number of people have labeled this stock *Porcellio scaber* not because of a detailed genitalia assessment, but because they assume it was collected in North America where it would likely be the only real option, but this bloodline was collected in Spain in 1997. A group of scientists who studied the genetics of color forms using introduced North American specimens of *P. scaber,* found their orange specimens "were always seen with a white edge" (Bhella et al. 2006). Their *P. scaber* had a wide, pale margin and over the course of the study they were unable to isolate a pure breeding line.

"Calico" This is the exact same bloodline as the Spanish "orange" but split from the parent stock in the late 1990s. The quality of the strain will depend on how recently it was separated from the parent stock since generations 1997-2013 had the ability to revert to normal color. Since then, they have become fully, or nearly true breeding but have become better and better looking (due to selection for color). Although a specimen's basic color does not change, full coloration is often not visible until the animal is mature or has reached full size. One unusual variation is clear in parts, so it is possible to see halfway through sections of the animal. A number of color varieties that are part of the calico variation have been bred for and sold under unique variety names.

"Dalmatian" These look similar to the original Dalmatians but usually have larger gray splotches and many tiny gray spots These were collected in 2012 from among a normal wild population in Michigan. Specimens do have two large lungs on each side of the pleopods. They immediately bred true which would be rather amazing if the females had been fully mature, since sperm can be retained for long periods.

Trachelipus rathkii
"Dalmatian" The original pied form was available in limited numbers in 2010 (first traded as *Porcellio* sp. in error). It is a beautiful white creature with black eyes and variable black spotting. I collected one specimen in northern Ohio in 2008 and it was a few years before there were a dozen. These look very much like *Porcellio scaber* but there are five white patches (lungs) along each side of the abdominal pleopods. Also, it behaves differently. It is prone to running quickly, jumping, and tucking in all the legs when disturbed so it is a bad idea to hold cage decorations above anything but the cage. It is a very beautiful form, but the stock has proven somewhat weak compared to the standard form and the number of specimens by 2013 only measured in the hundreds.

Armadillidium nasatum
"Peach Pillbug" This orange form is more of a peach color and retains the light and dark banding of the gray form. I found the first adult in woods behind my house in northern Ohio. I also collected one in northern Pennsylvania and another in eastern Virginia. Despite a good number of founding adults and work on getting and crossing hets since 2009, it took more than three years to produce a decent number of peach early instars. Adults will breed year-round, but immatures can take a year or more to reach maturity. *Armadillidium nasatum* is often listed as the largest species in the genus (21 mm versus 18 mm for *A. vulgare* but finding an old adult larger than 14 mm for either species is exceptional to the point of obscurity).

Porcellio laevis
"Halfwhite" stock was available from 2009-2014, but may have died out in captivity. 2009 offerings were not completely isolated and this species' variability by development is difficult to follow. Isolated, selected stocks threw mostly or entirely white, but the white has a gray blush or turns yellow variably. Occasional "spiny" specimens would pop up in halfwhite cultures. These are not terribly voracious but accept varying conditions and can be easily monitored since they are motile and

rarely burrow. I originally maintained two forms of this species from North America that proved vastly different in behavior and development. The first was acquired from mesquite scrub in the dry hot Sonoran Desert in Arizona circa 1998 and kept though 2006. Specimens were more resistant to desiccation than almost any other animal I have kept. They burrowed readily in substrate and would retract the legs and remain motionless. The largest specimens seldom reached 16 mm after many years and there was no difference seen in the male and female uropods. The second was the base stock for the Halfwhites and came from the same canyon in southern California as the "giant canyon" *P. dilatatus*. This habitat seems to offer selective pressure for large size since wild-caught specimens are often impressively large. Adults are the largest fully terrestrial isopod in North America and 20 mm in length is commonly achieved by specimens two year of age. Old males have long uropods that are double the length of the uropods of similarly sized females. This form rarely hides in the substrate with the legs retracted. **Recent Cultivars:** "Dairy Cow" is the name commonly used for the isolated pied form of *P. laevis* and this form can be nearly pure white with tiny black flecks. Adults rarely exceed 15 mm. The legs and body of this cultivar often have reddish highlights. "Orange" form is consistent in coloration and rarely exceed 15 mm. There is also a "milkback" form which is gray with an irregular pale stripe down the back.

Venezillo parvus
"Orange Micropill" Adults are usually 4-5 mm but can reach 7 mm. It is useful for a bio active setup but limited use for clean-up crews; slow reproduction and limited voraciousness. There are at least seven small native members of this genus and one or more adventives. They are only found in warm, southern areas of the country and habitats can be rather damp. These may prove useful for terraria unsuitable for the larger pillbugs but a female takes months to produce half a dozen mancae. They burrow and produce shallow tunnels but do not require much substrate. **Cultivars:** An orange form was isolated from specimens acquired with bumblebee millipedes collected in Miami Florida in 2004. It is a very common adventive in Florida but is so small it is rarely noticed.

Chapter 6. Select Species

Measurements given are those made using vernier calipers from the front of the head to the end of the telson. Due to the small size of isopods and the propensity for humans to make lousy estimates of small invertebrates, a set of vernier calipers is suggested for those interested in discovering or reporting the size of isopods. If you measure your biggest "monster" specimen you will be surprised when you find it does not measure as big as the maximum listed here and these are conservative compared to most size claims. 1" = 25.4 mm

Armadillidium corcyraeum **Uses:** Display and terrarium cleaner. **Size:** First matures and breeds readily at only 8 mm, but can grow to 15 mm. This Greek pillbug appears similar to the popular zebra pillbug, *A. maculatum,* but the pattern of white stripes and markings is a little different. A slightly dry area of the substrate is helpful but not required and excessive ventilation is not needed. Freshwater algae and carrots work better than dog food.

Armadillidium frontirostre **Uses:** Display. **Size:** 18 mm. Current stock was collected in Croatia and like every other gray *Armadillidium* it was hyped as the new giant species. It is relatively smooth and monotone, dark gray.

Armadillidium gestroi **Uses:** Display. **Size:** 18 mm. This French pillbug is another colorful, spotted species that may mimic the chemically protected pill millipedes (such as spotted forms of *Glomeris klugii* and *Glomeris oblongoguttata*). *Armadillidium* do not have chemical protection. Specimens often burrow and make shallow tunnels, so it can be difficult to find specimens even in a tiny cage. It is relatively hardy and breeds well.

Armadillidium granulatum **Uses:** Display, clean-up, and bio active media. **Size:** 20 mm. This is a widespread species across southern Europe, from Portugal in the west to Romania in the east and various Mediterranean islands in the south. Adult survival is more consistent than other *Armadillidium* (like *A. maculatum* and *A. nasatum*) where many adults live a year or so, but only a tiny percentage live three years or more. Adult *A. granulatum* regularly live two and a half to three years and seldom die prematurely. The color is patchy gray, usually with crayon yellow spots.

Armadillidium klugii Clown Pillbug **Uses:** Display. **Size:** Up to 18 mm. The common name references the multi-colored spots like a clown's outfit. Found in Albania, Croatia, Romania, and surrounding countries. The margin or skirt can be red, brown, or lacking. **Varieties:** Two common stocks are labeled after collection locations, "Dubrovnik" City in Croatia and "Montenegro," a country in the Balkans, both have yellow and white spots. A third is “Slano," named for a village in Croatia, which has only white spots. There is also a hybrid of “Montenegro” and *Armadillidium versicolor* traded as Twisted Clowns (McMonigle 2020).

Armadillidium maculatum Zebra Pillbug **Uses:** Display, clean-up, bio active media. **Size:** Available stock grows to a similar size of the following two species, around 14 mm, maximum size is 17 mm. This spectacular looking pillbug was included in the small guide (2013) only as a future possibility or dream species. I was able to get some, at what I considered then to be a very hefty price at the time, from an importer in 2013 (I had requested isopods from importers for decades and this was the first time the importer even knew what kind of bug I was talking about). It proved to be a very hardy and fast breeding species that was established quickly in culture. Ready availability of this beauty was one of the key components to the rapid establishment of an isopod hobby by early 2014. The stock originated in France, though the species also occurs in northern Italy. The normal pattern is dark with light margins similar to the warning colors of *Glomeris marginata* pill millipedes, though the bands are broken up variably in some specimens. None of the terrestrial isopods are known to have any chemical protection, but glomerids produce sticky, poisonous secretions when attacked. Most zebra pillbugs live 18-24 months, very rarely 36 months. **Cultivars:** The margins on the segments may be complete stripes or broken into spots. Either can be selected for over generations with reasonable consistency. Irregular forms can have mostly white coloration (commonly traded as "Dalmatian") or mostly gray which look a little like the standard coloration of *A. corcycreaum.* There is a "chocolate" form where the dark gray portions are brown and a form where the white margins are greenish-yellow.

Armadillidium versicolor Uses: Display, clean-up crew. **Size:** 9-12 mm. Adults resemble immature *A. vulgare* with excessive yellow spots and stripes. Full-grown specimens behave like immatures of larger *Armadillidium* because they hide in the top layer of the substrate, even at night. This species requires very little ventilation, does not need a dry area of substrate, and is sensitive to dry conditions. Tiny mancae are produced in as little as five months after the adults themselves are born.

Armadillidium vulgare **Uses:** Display and some clean-up for terraria. **Size:** 17 mm, few specimens measure above 12 mm. A friend of mine who owned a pet shop said he had a gift for me in 1995. I stopped in to find a clear 8" tub containing some recently imported "blue millipedes" and "pill millipedes" from Spain. The blue millipedes were small, gray *Ommatoiulus rutilans*, while the pill creatures were only *Armadillidium vulgare*. Although they were some of the largest *A. vulgare* I had ever seen I was disappointed by the rare pillipedes that turned out to be common pillbugs. Not one to look a gift horse in the mouth I took them home. They stayed in the small tub for a few years, were moved to a ten-gallon for a number of years and then were moved to a plastic shoebox. Large numbers died on the last move, but the problem was recognized, some holes were drilled in the lid, and the colony eventually rebounded. **Cultivars:** pied forms sold as "magic potion", orange, and albino forms have been culled from various collection locations.

Armadillo officinalis Squeaky Toy **Uses:** terrarium cleaner and display. **Size:** up to 18 mm. While other species rarely exceed 2.5 years from birth to death, this one lives up to nine years. It makes squeaky noises when handled but the sound is difficult to hear unless the creature is held next to the ear. They are hardy but prefer a dry area to retreat. **Cultivars:** a dark orange form is isolated.

Atlantoscia floridana Fast Dwarf **Uses:** clean-up. Like its larger cousin, this species scatters and runs rapidly when exposed. **Size:** It looks very similar to the common adventive *Philoscia muscorum,* but large, old 8 mm adults are only half the mass of a small, young adult *P. muscorum*. The two most significant differences are temperature and reproduction. They die if placed in a refrigerator and require tropical temperatures for survival while *P. muscorum* survive well below freezing. This species is highly productive and reproduces year-round in captive enclosures. The telson is shaped differently, there are three microscopic setae at the end of each antenna, and the body is narrow but the pattern of spotting is similar. However, unlike *P. muscorum*, this species so far has shown very little variation in the adult coloration.

Cubaris murina Little Sea **Uses:** Terrarium cleaner, cultivars for display. **Size:** 6-8 mm. The little sea pillbug is a common introduced species in Florida and

Texas. Man has transplanted it across the world's tropics and subtropics. The species name means mouse-colored (gray) but the common name was coined based on confusion with marine and small adult size. Females produce maybe a dozen young per brood, but can reproduce every few months. **Cultivars:** an isolated white form is available as well as a pale orange form called "papaya."

Cubaris sp. "rubber ducky" **Uses:** Display, expensive terrarium cleaner. **Size:** 12-14 mm. For many isopod hobbyists one of the key features is cuteness which is why this species, and a few relatives are the current rage. Vendors sell out of these at $25.00 for one tiny animal, but they are easy to keep and breed so the fad may see an abrupt ending. The rostrum is widened along the entire front of the head, resembling a duck bill. Combined with the small black eyes and yellow head the common name was born. If the yellow head and "bill" were not enough, the pattern on the pleon roughly looks like yellow tail feathers. When it walks the resemblance is lost because dark antennae extend out and move rapidly side to side (sort of like the nervous movement of wasp antennae). This species was originally collected in Thailand. Very little ventilation is required, and dry conditions are deadly. As with other *Cubaris,* a female can produce offspring every month or two, 6 to 24 per brood. The dark middle segment can be light in color which spawned the name "blonde rubber ducky" for some stock. The pale form could be from a different collection locale, but it is also a normal variation.

Cubaris sp. "red edge" **Uses:** Display, clean-up crew. **Size:** 11 mm. This species from Thailand was among the original species seen in the "*Cubaris* craze." Of course, many of these may prove to be different color forms of the same species or members of similar looking genera. It reproduces well and like other *Cubaris* the mancae are relatively few in number, but huge compared to the mother. A female only produces around a dozen young, but she can do this every month or two for her entire adult life. Specimens reach initial sexual maturity in as little as three months because they start out so large. **Cultivars:** There is a pure white stock and a fantastic looking orange form with red eyes that looks like the orange dango mushi (meaning 'dumpling bug' in Japanese) toy.

Cubaris sp. "red tiger" **Uses:** Display, possible terrarium cleaner. **Size:** 12 mm. This pillbug was among the early *Cubaris* imports from Thailand circa 2017. Specimens have a red skirt like the "red edge" but the body has uneven stripes that inspired the trade name. A striking relative with a related pattern and a white base color traded as *Cubaris* sp. "white tiger" showed up around the same time.

Cubaris sp. "Pak Chong" **Uses:** Display or cleaner. **Size:** 12 mm. The red head and tail and white racing stripes are unique. The common name is the name of the city in Thailand where specimens were originally collected. This is a very easy and hardy species that accepts high humidity and little ventilation or a little dryness. It eats a lot of dog food for something so small and fares poorly if underfed.

Cubaris sp. "panda king" **Uses:** Display. **Size:** 9 mm. Panda king is a highly reproductive, dry-loving creature collected from the mouth of a cave in Thailand. Very little ventilation is necessary, but some areas of the substrate should be dry. Adults are dark gray with a white middle section.

Cubaris sp. "Jupiter" **Uses:** Display. **Size:** 12 mm. Females are larger than males, lighter yellow and have darker markings on the body. This species is very sensitive to changes in parameters or neglect. Specimens feed well on tiny, high protein fish pellets.

"Cubaris dream" **Uses:** Display or cleaner. **Size:** 13 mm. This multi-colored species was collected in Viet Nam and looks a bit different in shape from identified *Cubaris*. It requires little ventilation and needs moisture but tends to hang on the dry side of the enclosure. The coloration is gray but there are red, yellow, and blue highlights that come together in a handsome pattern.

Cylisticus convexus Curly Isopod **Uses:** cleaner. **Size:** 14 mm. It readily rolls into a ball but it does not retract the antennae inside and the uropods are spiky, not rounded. Curly isopods are hardy and reproduce easily, but can take more than eight months to reach maturity. Pied and white forms so far seem to be caused by conditions or a virus and revert easily.

Haplophthalmus danicus Elongate Micropod **Uses:** Great for bioactive media since it is a very passive feeder, limited use for clean-up crews. **Size:** 3-4 mm. This widespread micro isopod was introduced to North America from Europe. Unlike other widespread adventives it is rarely noticed because it is tiny, and it is usually restricted to very damp habitat. It prefers rotten wood to most other foods. It seems to move very slowly because it is so small, but under magnification it appears to move quickly. A large colony of hundreds of individuals reproduces at a good pace, but a dozen specimens may not grow to a hundred in a year.

Nagurus cristatus Striped Dwarf **Uses:** Work well for most terraria but climbing propensity should be considered. **Size:** 6.5 mm. This is a unique looking species with its longitudinal stripes and white telson. The striped dwarf isopod is a surprisingly agile climber. If the sides of a glass or plastic enclosure have a thin layer of water or dirt, large numbers can be found hanging under the lid (care must be taken when popping the lid off). Specimens are able to survive the cold temperature 38° F (3° C) of a refrigerator for four days, but extended or deep cold should be avoided. This parthenogenetic stock, mostly likely single source, has been traded in the dart frog hobby since at least 2009. This species is commonly found in greenhouses and many tropical and warmer subtropical areas including central and western Florida.

Oniscus asellus Shiny Isopod **Uses:** Display and cleaner for damp terraria. **Size:** 17 mm. This very common species is usually the largest species seen in colder temperate areas of the US. It resembles *Porcellio,* but the outline is less streamlined, the flagellum is composed of three articles, and there are no visible lungs on the pleopods. **Cultivars:** A pied form is traded as "Mardis Gras" and orange forms under various names.

Philoscia muscorum Fast Isopod **Uses:** Limited. **Size:** Adults seldom exceed 10 mm in length. This is among the more common outdoor species introduced to

(Color Plate page 38)

8. *Porcellio magnificus* male and female dimorphism in uropods and antennae.
9. Adult male *Porcellio expansus* and *P. hoffmannseggi* compete for largest with *P. magnificus* (at one time a subspecies of *P. hoffmannseggi*).
10. Accidental hybrids nicknamed Twisted Clowns (McMonigle 2020).
11. *P. bolivari nickelsi* is a natural, pale orange species.
12. Male *A. klugii* "Montenegro" guarding a female *A. versicolor* (these are the parents of hybrids in figure 10).
13. *Tuberillo borneanus* mature specimens.
14. Spiky *Porcellio* from the Canary Islands.
15. *Cubaris* sp. "red edge" from Thailand.
16. Gold Dalmatian *P. scaber,* Michigan origin. These display some blue fluorescence under 356NM black light.

Porcellio laevis "dairy cow"

Porcellio flavomarginatus

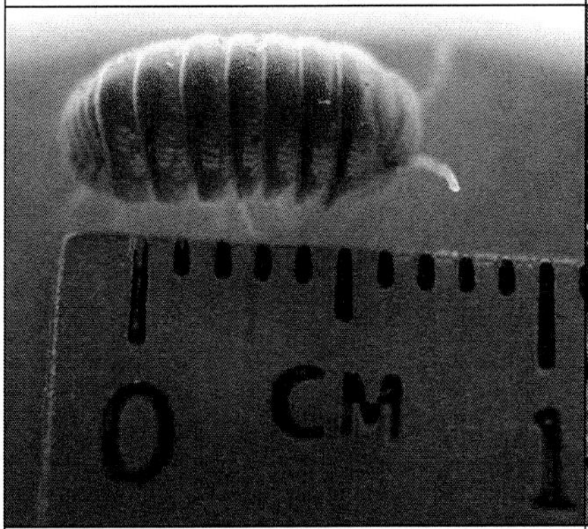

Venezillo arizonicus colleted in Arizona 2009 (scale markings are millimeters.) Photo by Peter Clausen.

Armadillidium corcyraeum

8

9

10

11

12

13

14

15

16

North America from Europe. The name refers to its rather impressive land speed. The body color is normally offset by a contrasting margin of red or white. A wide array of colors is common in groups of wild-caught adults but the colors are not nearly so bright nor as pretty as they are for cultivars of other species. It is easy to keep specimens alive and healthy in captivity for two years, but captive production of immatures is rare.

Porcellio bolivari Yellow Cave Isopod **Uses:** Display. **Size:** Old males can obtain a body length as great as 25 mm, but less than 20 mm is common. Males have significantly longer uropods (up to 7 mm) than females (less than 2 mm) and average larger. All specimens in captivity might be from the original group found on the Bernia rock faces in Spain (McMonigle 2018). Specimens found in caves in the Penyal d'Ifac park in Calp look similar, but the body color is a darker, rusty yellow (D. Garcia). Penyal d'Ifac park's location is a few miles east of the Bernia Mountains, on a huge limestone outcropping in the sea. There is a subspecies (*P. bolivari nicklesi*) that is peach to orange in color which lacks the yellow stripes. It also lacks the black markings that are located on the outer, rear edges of the pleurites of the nominal variety. Specimens of *P. b. nicklesi* have also made it to captive culture. Both subspecies do best with good ventilation.

Porcellio canariensis Canary Island Spiky **Uses:** Display, but small and cryptic. **Size:** 8 mm. This stock was found on the Canary Islands and specimens are covered with spiky tubercles. A good photo shows rows of impressive spikes but a large specimen is small and appears fuzzy when not magnified. They primarily eat dead leaves and dried grasses, not dog food, but carrots are a favorite food.

Porcellio dilatatus "Giant Canyon" **Uses:** cleaner for large enclosures. **Size:** 19 mm. Although not a rare species, those traded in the hobby since 2007 were from a single source, a canyon in southern California. It is a large, hardy, reproductive stock that is resistant to dryness and survives high humidity. It can take several years for large specimens to show up in a colony, but they can be very impressive, bulky creatures. Medium to large specimens are usually a very monochromatic gray but large, old specimens can have red-brown edges. It is a burrowing species with specimens often found an inch deep in the substrate. It has a highly pronounced curl defense and often stays that way longer than I care to check.

Porcellio expansus Yellow Dragon **Uses:** Display. **Size:** 26 mm. Male uropods to 13 mm. For isopod enthusiasts, this is the dream species and may be the most sought after terrestrial isopod. It is also a little difficult. While fads of tiny pills and color morphs ebb and flow, the appeal of this species is timeless. It is impressive in size, coloration, and body structure. The expansive margins, variable white checkers on the body and striped antennae create a unique and beautiful look. Two geographic forms are kept, one with yellow margins and another with a wide orange band inside the yellow. The yellow margins have a narrow, orange stripe along the rear edges. "Yellow Dragon" are among the most strongly sexually dimorphic of all the terrestrial species. Males and females are similar in coloration and body length, but the appendages of the male are exaggerated. Males develop very impressive uropods that grow more exaggerated in old age. They have much wider and longer antennae along with a wider head to hold them. Though a less noticeable feature, the front three legs of the male are enlarged. Large adult males have a pronounced ridge of serrated "scales" across the front three tergites and less notable raised tubercles on the head and other segments. The largest row of teeth on the first pereonite is used by males during aggressive shoving matches over a receptive female. They are used like the horns of fighting beetles. Fish food pellets, crab slices, and shrimp tails are eaten with gusto. Small, crustose lichen is chewed off bark with speed, but bark, wood, and leaves generate little interest. Cuttlebone is eaten, but it takes some time for a small chunk to be worn down. Small pieces of carrot are chewed on, but never finished. **Cultivars**: An orange form (where the gray center area is orange, not the inner margins) exists but is not yet isolated.

Porcellio flavomarginatus **Uses:** Display. **Size:** 15 mm. This handsome, black and white species hails from Greece, Crete, and the surrounding islands. It stands up on its legs and adults are arboreal in nature. Specimens like to climb and hang beneath bark angled against the side of the enclosure. They are prone to dying from limited ventilation. Males and females are difficult to tell apart from above. Very young specimens almost look like a negative image of the adult, stay in the substrate and require higher moisture. They grow rapidly and can mature in just three months.

Porcellio haasi **Uses:** Display. **Size:** 18-24 mm, male uropods up to 6 mm. This handsome species from Spain is unique for the large yellow rectangles along the back. However, one of the commonly traded forms ("dark") has broken speckles similar to the pattern of *Oniscus* and another looks like the Almeria form of *P. ornatus*. The "high yellow" form has large solid squares while the "giant" form usually has smaller yellow squares. All of the forms have a white margin (skirt) around the body. So far, none of the other giant species have small and large forms. Adult males have notably long uropods. Specimens require good ventilation and hang out in the dry areas of the cage.

Porcellio hoffmannseggi Titan Isopod **Uses:** Display **Size:** 28 mm. Male uropods up to 8 mm. This is a massive and impressive species native to Portugal and Spain (mainland and Baleric Islands). Old adult males can be long and narrow. This is the only giant Spanish species that survives stagnant air for extended periods and can survive a range of conditions almost as expansive as *P. scaber*. Specimens aggressively feed on pelleted foods for fish, other fish meal products and slices of imitation crab. Crustose lichen is chewed off of bark, but the bark is hardly touched. Small mosses are often consumed readily. Wood and leaves are consumed slowly. The female begins to bear young at around 18 mm and nine months of age. Reproduction is common, but if there is significant overcrowding or limited hiding areas the mancae disappear into hungry maws. Shredded leaves, small flat pieces of bark, and branchy moss can provide adequate shelter for mancae. Once they make it past the first molt the survival rate is very good. Titan isopods live at least two years and adults rarely die in the first twenty-four months. **Varieties:** A variety from a different collection locale with darker gray coloration and thinner white margins is traded as "black." Pied forms have been documented, but may not be isolated at this time.

Porcellio magnificus **Uses:** Display **Size:** largest *Porcellio* reaching 32 mm in body length in nature but almost never over 26 mm in culture. This massive Spanish species is naturally orange in color. Small specimens are easy to differentiate from orange *P. scaber* adults because there is a narrow white margin around the body, and the legs and tips of the antennae are white. Adult males have very long uropods and it is easy to differentiate males and females when the specimens are only a centimeter long. Magnificus is one of the slowest growing *Porcellio* species, taking 12-16 months to reach initial maturity and twice as long to reach full size. Good ventilation is required.

Porcellio ornatus Ornate Isopod **Uses:** Display. **Size:** Old specimens can become pretty large creatures that are often 22 mm from head to telson. They usually meet or exceed the normal maximum length of the largest terrestrial species encountered in North America (European adventive *Porcellio* and *Oniscus*). The "high yellow" and Almerian "white skirt" forms require good ventilation and moisture, so they are tricky for most keepers. **Varieties:** yellow dot, Almeria, and high yellow are wild color forms.

Porcellio spinicornis Brickwork **Uses:** Display. **Size:** 15 mm This European adventive is widespread in North America around cement structures. Brickwork is a great test species for keepers learning to keep the big Iberian species. It can die from dampness but cannot be too dry. Brickwork can be confused with other *Porcellio* but the head is usually darker than the body and there is a nearly microscopic spine at the end of the third antennal segment. Specimen color can change to match background after molts.

Porcellio succinctus White dragon **Uses:** Display. **Size:** Males reach 25 mm in body length, with uropods up to 11 mm. Uropods of older females can reach 6 mm. This large *Porcellio* from southeastern Spain is uncommon in nature and rare in captivity. It resembles *P. expansus*, but with a less flattened profile and more narrow margins (skirt). Likewise it is gigantic and the males have extremely long uropods which become more extreme with age, but the males are much less tuberculated. There is only a small row of "scales" (tubercles) across the front pereonite of older males (similar to *P. bolivari*). Good ventilation, accessible moisture, and warm temperatures above 72° F (22° C) are preferred. Specimens rarely feed on leaves but feed well on fish food pellets. Specimens can live three years in captivity including the nine to twelve months needed to reach maturity. The female produce 30 or more young per brood, but usually only once and the survival rate can be less than 15%.

Porcellio werneri Greek Shield **Uses:** Display. **Size:** ~ 20 mm. This handsome Greek species is almost as wide as tall, like a shield. It is similar in shape and size to the flattened, gray *Porcellio spatulatus*, but with a nearly black body and bright white margins. Males have wider margins and can be identified on sight. Specimens cling to the upper surface of bark and are more likely to run than play dead. They do not require excessive ventilation but stay on the dry side of the enclosure and will die if unable to get away from wet substrate.

Porcellionides pruinosus Powder blue. **Uses:** cleaners. **Size:** These reach about 10 mm and are rather quick growing and easy to care for. One interesting aspect is the black and white banding of the antennae. Sometimes the uropods are orange. The body is gray but has a powdery look from different angles. The exoskeleton is pliable and does not seem to crunch like most species when accidentally damaged. **Cultivars:** a pied form, an orange form, and a white form are readily available.

Trichorhina biocellata Jungle Micropod **Uses:** cryptic cleaner. **Size:** ~3 mml. Older immatures and adults can be purplish gray, especially from a distance, giving rise to the name "purple micros" on many sales pages. Many specimens are orange. Commonly burrows and holds onto dirt rather than the underside of wood. The stock was collected in Costa Rica in 2007.

Trichorhina tomentosa White Micropod **Uses:** aggressive cleaner. **Size:** Old adult specimens rarely exceed 4 mm in body length. This species is native to Central and South America but was long ago introduced to Florida and is commonly found in greenhouses worldwide. If a culture source cannot be found, a trip to a few local greenhouses will likely yield results. I originally acquired specimens in 1999 and since that time they have become one of the most common and desired species for use in dart frog terraria. This species burrows (usually less than half an inch deep). Although a female only has half a dozen young at a time, there are no specimens in a culture that cannot have young (reproduction is through parthenogenesis).

Tuberillo borneanus Borneo Alligator **Uses:** display or cleaner. **Size:** 6 mm. This bumpy, volvating species is a recent arrival and newly described species (2010). Females produce a dozen or more large offspring at a time. It does great on freshwater hair algae and eats most isopod fares. High humidity and low ventilation work well.

Glossary

Adventive: Naturalized or introduced species.

Biramous: Dividing into two main trunks or branches.

Bunching: Clustering together during dry periods to reduce water loss.

Desiccate: To dry up or dry out; dehydrate to the point of death.

Dimorphic: Two distinct shapes or forms.

Flagellum: The terminal segment of the isopod antenna normally composed of two to twelve articles, the number of articles can be useful in identification. There are no muscles in this section of the antenna.

Het: Short for heterozygous, refers to an animal with one dominant and one recessive gene for a specific trait. Hets look the same as an animal with two dominant genes. If crossed with another het the theoretical offspring will be 50% het, 25% dominant, and 25% recessive.

Hypopus: A unique stage of certain types of mites that is a response to reduced food, they molt into a suction cup stage and search out exoskeletons of living arthropods to cling to.

Iridovirus: Here: Isopod iridescence virus, it results in blue coloration of a percentage of adult isopods but has always proven lethal within a few months.

Littoral: Of or living within the area between high and low tide.

Manca (pl. **Mancae**): 1st instar isopod.

Marsupium: Specialized egg pouch between the front legs and comprised of oostegites.

Maxilliped: Outer, third pair of mouthparts that arise from the thorax .

Oostegites: Plates that form the marsupium and arise from near the base of the pereopods.

Parthenogenesis: Development through unfertilized eggs.

Pereon: Middle section of the isopod body, thorax as separate from the head with incorporated maxilliped segment and the pleon.

Pereopods: Ventral uniramous appendages attached to the pereon. These are the walking legs for most species though they can be useless for locomotion and adapted for filter feeding or prey capture in some marine isopods.

Phenotype: The appearance of a specific dominant trait such as gray body color. An animal that is heterozygous will appear the same as one with both dominant genes.

Pleon: Abdomen as separate from the pereon, contains the pleopods, telson, and uropods.

Pleopods: Biramous appendages on the pleon that form the gills and can be adapted in relation to mating for the males, termed swimmerets for many aquatics.

Telson: Terminal body segment or tail. This is part of the pleon and may be called a pleotelson to differentiate it from the telson of a scorpion.

Uniramous: Nonbranching.

Uropods: Biramous or uniramous appendages on either site of the telson; stylets. These form the paddles of the tail in swimming isopods, lobsters, etc.

Bibliography

Bhella, S., Fung, E., Harrison, J., Ing, B., Larsen, E. Selby, R.D. (2006) Genetics of Pigmentation in *Porcellio scaber* Latreille, 1804 (Isopoda, Oniscidea) --*Crustaceana* 79 (8): 897-912

Caldwell, Robert (2008) Giant Deep-Sea Isopod (Bathynomus giganteus) -- *Invertebrates-Magazine*. Vol. 7, Issue 4. pp. 19-20.

Dunn, Gary A. Editor (1993) *Caring for Insect Livestock: An Insect Rearing Manual.* Special Publication No. 8, Young Entomologist's Society, Lansing, MI.

Forster, R. R. & L. M. (1980) *Small Land Animals of New Zealand.* John McIndoe Limited. Dunedin, New Zealand.

Fritzsche, Ingo (2013) Auf Wirbellosenpirsch in der Levante - Spannende Entdeckungen an der Sonnenkuske Spaniens. - *Bugs Das Wirbellosenemagazin.* Nr. 1,1. p.32-36.

Hellweg, Michael R. (2009) *Raising Live Foods.* TFH Publications Inc. Neptune City, NJ.

Jass, Joan & Klausmeier, Barbara (2001) *Terrestrial Isopod (Crustacea: Isopoda) Atlas for Canada, Alaska, and the Contiguous United States.* Contributions in

Biology and Geology. Milwaukee Public Museum.

Linsenmair, Karl E. (2007) Sociobiology of Terrestrial Isopods in: Evolutionary Ecology of Social and Sexual Systems: Crustaceans as Model Organisms (J. Emmett Duffy and Martin Thiel).

Kneidel, Sally (1994) *Pet Bugs A Kid's Guid to Catching and Keeping Touchable Insects*. John Wiley & Sons, Inc. New York.

Masters, Charles O. (1975) *Encyclopedia of Live Foods*. T.F.H. Publications, Neptune City, NJ.

McCormick, Stephanie (2008) Arthopoda 2: Subclass Malacostraca. Biology 377.

McMonigle, Orin (2001) *Assassins, Waterscorpions & Other True Bugs Care and Culture*. Elytra and Antenna, Brunswick, OH.

McMonigle, Orin (2003) *Giant Centipedes The Enthusiast's Handbook*. Elytra and Antenna, Brunswick, OH.

McMonigle, Orin (2004) Terrestrial Isopods. -*Invertebrates-Magazine*. September, 4:4 pp. 9-12.

McMonigle, Orin (2011) *Invertebrates for Exhibition Insects, Arachnids, and Other Invertebrates Suitable for Display in Classrooms, Museums, and Insect Zoos*. Coachwhip Publications, Landisville, PA.

McMonigle, Orin (2013) The Beauty of Volvation and the Pretty Peach Pillbugs. -*Invertebrates-Magazine*. June. 12:2. pp.6-7,12-13.

McMonigle, Orin (2017) Current Account and Captive Considerations for the Ornate Isopod Porcellio ornatus Milne-Edwards, 1840. -*Invertebrates-Magazine*. August. 16:3 pp. 12-14.

McMonigle, Orin (2018) Biography and Biology of *Porcellio bolivari* Dollfus, 1892 (Isopoda: Oniscidea) the Yellow Cave Isopod. -*Invertebrates-Magazine*. August. 17:3 pp. 16-17.

McMonigle, Orin (2020) Twisted Clowns: A Story of Love and Loss (or) Terrestrial Isopod Hybridization. -*Invertebrates-Magazine*. August. 19:3 pp. 7-9.

Mocquard, J.P., Pavese, A. & Juchault, P. (1980) Seasonal reproduction determinism of *Armadillidium vulgare* females (Crustacea, Isopoda, Oniscoidea): 1. Action of temperature and photoperiod. Ann. --*Sci. Nat. Zool. Biol.* Anim. 2, p. 91-98.

Schmalfuss H. (2003) *World catalog of terrestrial isopods (Isopoda: Oniscidea)*. Stuttgarter Beiträge zur Naturkunde, Serie A (Biologie), Bd. 654, S. 1–341, Stuttgart.

Scott L. Kight, Michele Martinez, Aleksey Merkulov (2001) Body Size and Survivorship in Overwintering Populations of *Porcellio laevis* (Isopoda: Oniscidea)-- *Ent. News* 112(2): 112-118.

Souty-Grosset, C., Bouchon, D., Mocquard, J.P. & Juchault, P. (1994) Interpopulation variability of the seasonal reproduction in the terrestrial isopod *Armadillidium vulgare* Latr.(Crustacea, Oniscidea): A review. -- *Act. Oeco.* 15, p. 79 -91.

Werner, F. Ph. D. & Olson, C. M.S. (1994) *Insects of the Southwest*. Fisher Books, Tucson, AZ.